T0176005

Mysteries of the Deep

Mysteries of the Deep

How Seafloor Drilling Expeditions Revolutionized
Our Understanding of Earth History

James Lawrence Powell

The MIT Press

Cambridge, Massachusetts | London, England

The MIT Press would like to thank the anonymous peer reviewers who provided comments on drafts of this book. The generous work of academic experts is essential for establishing the authority and quality of our publications. We acknowledge with gratitude the contributions of these otherwise uncredited readers.

This book was set in Stone Serif and Stone Sans by Westchester Publishing Services. Printed and bound in the United States of America.

Library of Congress Cataloging-in-Publication Data

Names: Powell, James Lawrence, 1936– author.
Title: Mysteries of the deep : how seafloor drilling expeditions revolutionized our understanding of earth history / James Lawrence Powell.
Description: Cambridge, Massachusetts : The MIT Press, 2024. | Includes bibliographical references and index.
Identifiers: LCCN 2023017995 (print) | LCCN 2023017996 (ebook) | ISBN 9780262048927 (hardcover) | ISBN 9780262378208 (epub) | ISBN 9780262378215 (pdf)
Subjects: LCSH: Submarine geology. | Oceanography. | Ocean bottom. | Geology—Research.
Classification: LCC QE39 .P69 2024 (print) | LCC QE39 (ebook) | DDC 551.46/8—dc23/eng/20231016
LC record available at https://lccn.loc.gov/2023017995
LC ebook record available at https://lccn.loc.gov/2023017996

10   9   8   7   6   5   4   3   2   1

In memory of Peter Molnar (1943–2022)
Graduate of Oberlin College and Winner of the
Crafoord Prize in Geosciences

To unpathed waters, undreamed shores . . .
     —William Shakespeare, *The Winter's Tale*

Contents

x   Contents

1
Introduction

As the 1960s began, the only scientific expedition to have studied the world's oceans had taken place 90 years before, near the end of the age of sail. Between 1872 and 1876, HMS *Challenger* traveled 130,000 km around the globe, measuring the depth and temperature of the oceans' waters. The scientists aboard studied the soft sediment that covered the seafloor and the hitherto unseen creatures their dredge brought up.

This revolutionary voyage launched the science of oceanography but only scratched the surface of the seafloor. What lay beneath the sediment cover remained unknown. The paucity of depth measurements compared to the vastness of the ocean floor allowed only the vaguest idea of seafloor topography. Geologists were left with the impossible task of trying to understand Earth while having access to only the 30 percent exposed as continents. It is thus no surprise that even by the mid-twentieth century, many fundamental scientific questions about Earth remained unanswered—but that's not all. For most of the first half of the twentieth century, the field of geology had operated without an overarching theory of how Earth works.

Nineteenth-century geologists had two theories that eventually fell out of favor. *Permanence* was the view that the continents and ocean basins have always been where we find them today and have retained their original size and shape. The two geological features were interchangeable, as continents had foundered to become ocean basins, which in turn had been uplifted to become continents. The second theory, *contraction*, explained the motive force that powered Earth: Over geologic time, as it lost its original heat, Earth cooled and shrank. This forced the rigid crust to contract and wrinkle to conform to the shrinking interior, analogous to the skin of a drying apple. The wrinkles were supposed to correspond to our mountain ranges.

As the nineteenth century ended, new discoveries brought an end to both permanence and contraction theories. The "basement" rocks of the ocean floor—those beneath the thin cover of sediment—turned out to be unlike those of the continents and not interchangeable with them. Mountains were found in effect to float on a fluid interior layer in the way an iceberg floats in the ocean: they are buoyant and cannot sink to form ocean basins. Then, shortly after the discovery of radioactivity in 1893, Pierre Curie found that each radioactive decay event releases a small amount of heat. Multiply that by Earth's innumerable radioactive atoms and our planet is as likely to be heating up as cooling down.

Thus, as the new century began, geology found itself a science without a theory. Then in 1912 came Alfred Wegener with his theory that continents move, or drift, across Earth's surface—the opposite of permanence. Continental drift explained a wide range of facts about Earth, including the origin of mountains. But it confronted an ongoing debate among geologists as to whether Earth operates only slowly or rapidly and at times catastrophically, which moving and colliding continents would surely exemplify. Geologists had named the gradual process theory *uniformitarianism*. It came to be so universally accepted as to be regarded as essentially proven. One of the first to question the logic of uniformitarianism was a young Stephen Jay Gould, in his first scientific paper written in 1965 and titled "Is Uniformitarianism Necessary?"[1] Typically, scientists do not reject a theory until they have a better one to replace it. But so strong was the dominance of uniformitarianism that geologists were willing to reject continental drift despite having no substitute.

Thus, as of the mid-twentieth century, the lack of a ruling paradigm left geologists unable to answer several fundamental questions about Earth:

- If contraction did not cause mountain ranges, what did? Why are many of them, such as the Andes, Appalachians, and Himalayas located near the margins of continents? Geologists had deduced that the forces that had pushed up the Appalachians, for example, had come from the southeast, but in that direction lies only open ocean, which could not push anything.

· What was the origin of the steep-sided volcanoes that make up the Pacific Ring of Fire, such as Mount Fuji and the Cascade Range of the Pacific Northwest?

· Mapping during World War II had revealed that the topography of the seafloor differs in nearly every respect from that of the continents. One example is that a lofty submarine mountain range runs down the center of the Atlantic Ocean floor. Paradoxically, the deepest ocean water is not near the center of the oceans, but in the trenches that lie around the margins of the ocean basins. What caused this?

· Why do the coastlines of Africa and South America appear to fit together like two pieces of a jigsaw puzzle, with the Mid-Atlantic Ridge in between?

· Why, at several places in the Pacific Ocean, notably the Hawaiʻian Islands, are volcanic islands arranged in a long chain, with those at one end (Hawaiʻi) the youngest and those at the opposite end (Nihoa) the oldest and most eroded?

Another set of unanswered questions had to do with ancient climates and life under the sea:

· What caused the ice ages, when huge continental-scale glaciers advanced from the poles and retreated, not once but many times? Could the repeating cycles of Earth's position and orientation in space have affected global temperatures and caused the ice ages?

- Antarctica is now ice covered but once was warm and tropical. How long ago did Antarctica glaciate and how much would sea level rise were its ice to melt today due to global warming?
- *Challenger* found that life is present even in the deepest ocean waters. In the 1930s, scientists had found evidence that microbial life exists in sediments beneath the seafloor. Could it be widespread on Earth, and possibly even on other planets like Mars?

These were among the unanswered questions of geology as of the 1950s. Yet in less than one generation, a single research program had at least partially answered each of them, revolutionizing the science of geology.

2
The Voyage of HMS *Challenger*

On December 7, 1872, a new type of oceangoing vessel put out from the British port of Sheerness in northwest Kent, bound for the four corners of the world. As shown in figure 2.1, HMS *Challenger* was a three-masted corvette, the smallest of Britannia's warships at a time when the empire ruled the waves. *Challenger*'s purpose was not war, however, but scientific exploration. For the next four years, it would circumnavigate the globe collecting scientific data on the world's oceans and retrieving thousands of specimens from the ocean floor, and, in the process, inventing the science of oceanography.

Era of Exploration

The nineteenth century was an era of scientific expeditions. Several voyages of exploration had preceded *Challenger*, most famously the 1831–1836 voyage of HMS *Beagle* with a young Charles Darwin aboard. These shipboard scientists came along to study the flora, fauna,

Figure 2.1
HMS *Challenger* off the Kerguelen Islands in the southern Indian
Ocean. Hand-colored woodcut.
Source: North Wind Picture Archives / Alamy Stock Photo.

and peoples of distant lands, rather than the water that
had carried them and what lay beneath it. That would
be the principal mission of the *Challenger* expedition.

The introduction to the report of the *Challenger* expedition reviewed the exploratory voyages that had preceded it, noting that they had proved that life existed
"at vast depths in the sea and that, with a little care, this
life could be investigated by ordinary and well-known
means." But these earlier voyages had been restricted
to coastal waters and enclosed seas, leaving "the vast
ocean . . . scientifically unexplored." This opened the way

for a great exploring expedition that "should circumnavigate the globe, probe the most profound abysses of the ocean, and extract from them some sign of what went on at the greatest depths."[1]

The Royal Society of London set out the objectives of the voyage more precisely:

> To investigate the physical conditions of the deep sea in the great ocean basins (as far as the neighborhood of the Great Southern Ice Barrier) in regard to depth, temperature, circulation, specific gravity and penetration of light.
>
> To determine the chemical composition of seawater at various depths from the surface to the bottom, the organic matter in solution and the particles in suspension.
>
> To ascertain the physical and chemical character of deep-sea deposits and the sources of these deposits.
>
> To investigate the distribution of organic life at different depths and on the deep seafloor.[2]

Only 60 m long and 12 m wide, *Challenger* packed in a crew of 269 men, one-quarter of whom would eventually desert, disillusioned and wearied by the tedium of the repetitive tasks assigned to them by the "scientifics" aboard, as the seamen called their educated shipmates.

The most tedious task was the daily lowering of a dredge to the seabed, which could take three hours to reach 2,000 fathoms (1 fathom is about 1.8 m)—and they would reach much greater depths than that. Then after the dredge had been towed behind the ship for a while, it had to be hauled up—thankfully with the

Figure 2.2
The officers and scientists of HMS *Challenger*, 1874. In the front row
wearing a white suit is Sir Charles Wyville Thomson, chief scientist.
Source: Wikimedia Commons, June 23, 2008, https://commons.wiki
media.org/wiki/File:Challenger_expedition.jpg.

assistance of a small engine. To the untrained eye, the
recovered dredge held not some treasure of the deep,
but typically the same monotonous mud, varying only
by its two colors: white "ooze" and red clay. As Lord
George Campbell wrote of the former, "The mud! Ye
gods. Imagine a cart full of whitish mud, filled with
minutest shells, poured all wet and sticky and slimy
on to some clean planks, and then you may have some
faint idea of how *Globigerina* [one of the most common
species of microscopic plankton] mud appears to us.
In this the naturalists paddle and wade about, putting

spadefuls into successively finer and finer sieves, till nothing remains but the minute shells, &c."[3]

Another important task was to drop a weight to the seafloor to measure ocean depth. This "sounding" line was marked with flags every 25 fathoms. These were counted as the line played out, so that the measurement was accurate only to that extent, leaving smaller bottom features unrecorded. The scientists had made the reasonable assumption that the seawater would cool with depth, and for a while it did, but then at some lower level it warmed. This was an important scientific finding in that it suggested that large masses of seawater could exist somewhat independently of each other. To measure the water temperature at different depths, the crew used a special thermometer. Flasks to sample the water were lowered and retrieved. Most of these instruments were invented for the voyage. As shown in figure 2.3, *Challenger* sailed 70,000 nautical miles. The crew conducted nearly 500 deep-sea soundings and 133 bottom dredges, discovering 4,700 new marine species. The scientific report of the voyage, prepared by assistant scientist and naturalist John Murray, took 20 years to complete and comprised 50 volumes.[4]

Before *Challenger*, scientists knew far less about the seafloor than they did about the surface of the airless Moon, so clearly visible through telescopes. Based on a small number of soundings, scientists before *Challenger* had thought that beyond the continental shelf (the gently descending section of seafloor between the coastline and

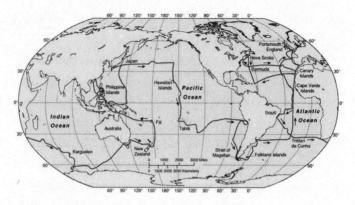

Figure 2.3
The route of HMS *Challenger* from 1872 to 1876.
Source: Peggy Musto, Challenger 150: The HMS *Challenger* Anniversary Program, August 9, 2014, https://www.slideserve.com/peggy/challenger-expedition-1872-1876.

the steeper continental slope), the seafloor was a featureless plain covered by sediments eroded from the adjacent land. This picture had already begun to change when in 1853 US Navy officer Matthew Fontaine Maury published maps based on 200 soundings in the North Atlantic Ocean. The sounding method had used a weighted hemp rope, which had the limitation that when pulled sideways by currents and stretched by its own weight, it gave inaccurate results. Still, the soundings were good enough to suggest that the seafloor in the middle of the North Atlantic rises higher than on either side, a feature Maury named the "Dolphin Rise" in honor of USS *Dolphin*, from which he made the soundings.

Setting Sail

Challenger docked at Portsmouth to load stores and on December 21, 1872, sailed for Lisbon, where inclement weather delayed the voyage for three weeks, after which the ship departed for the British territory of Gibraltar. Not far out of Lisbon, in 2000 fathoms of water, the dredge brought up an unusual, and for some, unexpected, specimen. It was a sea lily, or crinoid, of the echinoderm phylum, which also includes starfish, sea urchins, and the like. The sea lily, as the name suggests, looks like a flower sitting atop a long stalk—but it is an invertebrate animal. The specimen came from far below the limit at which life had been thought to exist on the seafloor, according to the prevailing azoic ("without life") hypothesis of marine scientist Edward Forbes (1815–1854). He had participated in an 1841 dredging voyage in the eastern Mediterranean aboard HMS *Beacon*, which made over 100 hauls down to 230 fathoms.[5] Forbes found that "the number of species and of individuals diminishes as we descend," and extrapolated to conclude that below around 300 fathoms, the number would reach zero and life would no longer be possible.[6] The hypothesis seemed reasonable, for how could living creatures withstand the enormous pressures, deep cold, and perpetual darkness of the abyss? How could they survive without photosynthesis? The famous nineteenth-century naturalist and geologist Louis Agassiz agreed, writing that "the depths of the ocean are quite as impassable for marine species as high mountains are for terrestrial animals."[7]

The azoic hypothesis became so entrenched that it persisted despite having been clearly disproven by the presence of many living specimens found at greater depths than Forbes's limit. In 1818, for example, the British ship *Isabella*, on its way to search for the fabled Northwest Passage, dredged up a starfish in Baffin Bay from a depth of about 600 fathoms. From 1868 to 1870, HMS *Porcupine*, with Scottish natural historian and marine biologist Charles Wyville Thomson (1830–1882) aboard—he would become chief scientist of the *Challenger* expedition—dredged in deep water off Great Britain, Spain, and in the Mediterranean, finding hundreds of new species representing all the invertebrate groups down to at least 650 fathoms.

The refusal to accept that life existed at greater than 300 fathoms, despite overwhelming evidence to the contrary, became a classic example of how scientists can continue to accept a hypothesis despite falsifying evidence against it.[8] But if a hypothesis is false, the evidence will eventually become too abundant to deny, though its acceptance may sometimes require a new generation of scientists. Nevertheless, even a false hypothesis may advance science, as the great biologist Thomas Huxley explained in 1853: ". . . There are periods in the history of every science when a false hypothesis is not only better than none at all, but it is a necessary forerunner of, and preparation for, the true one."[9] The azoic hypothesis does not really measure up to that description; it was simply wrongheaded from the outset and led nowhere.

At the time of the *Challenger* expedition, scientists and the educated public alike were fascinated by Darwin's theory of evolution—his "descent with modification." The continual change of terrestrial ecosystems required animals to change in response, to adapt, Darwin postulated, and only some were able to do so and produce offspring. But in the dark and unchanging realm of the deep ocean, as scientists perceived it, there would be little to no pressure to adapt—organisms found only as fossils on land might thrive as "living fossils" in the depths.

Huxley, who so vigorously championed the theory of evolution that he became known as Darwin's Bulldog, subscribed to this view, writing that "it may be confidently assumed that the things brought up [by *Challenger*] will be . . . zoological antiquities, which in the tranquil and little-changed depths of the ocean have escaped the causes of destruction at work in the shallows, and represent the predominant population of a past age."[10] Agassiz agreed, writing that "in deeper waters we should expect to find representatives of earlier geological periods."[11] But this did not mean that he accepted Darwin's theory. Indeed, Agassiz became one of its strongest opponents, writing: "There is in all this nothing which warrants the conclusion that any of the animals now living are lineal descendants of those of earlier ages."

Among the more than 4,000 new species that *Challenger* discovered were some that had previously been thought to be extinct, but nothing to match the description of "living fossil" ever came up in the dredge. As Loren Eiseley wrote in *The Immense Journey*, "These were only such discoveries as might be expected when any vast unexplored region is first investigated, whether it be land or sea."[12] This was a special disappointment for Wyville Thomson, who with each dredge haul "eagerly looked out" but in vain for a trilobite, whose last representatives disappeared in the great Permian-Triassic mass extinction 250 million years ago.[13]

To jump ahead for a moment, later in the voyage Wyville Thomson did discover an organism at least as interesting as a living fossil: a so-called "missing link." Darwin had followed his mentor Charles Lyell in the belief that geological change is slow and imperceptible. It follows that evolution mostly proceeds by the slow, steady accumulation of favorable traits, rather than being punctuated by sudden leaps. Scientists should find fossil species that are transitional between an ancestor and a descendant—but they seldom do. Darwin put this scarcity down to gaps in the geological record, periods when rocks had been eroded away or never deposited, hiding the links between the before and after. One reason for the great public interest in the *Challenger* voyage was that, by searching in previously unexplored areas, the scientifics might well come across such shadowy creatures—and so they did.

In the last few days before the ship left Australia in June 1874, Wyville Thomson led a collecting expedition in Queensland and returned with specimens of an exceedingly odd fish that had both lungs and gills, and not only fins but primitive legs used for walking on land. The lungfish belong to the Sarcopterygii, the group that gave rise in the Devonian period (420–360 million years ago) to land-dwelling vertebrates called Tetrapods. Thus, the lungfish and its relatives played a pivotal role in the movement of life from the sea to the land. Today, only three genera of lungfish survive, each found only on a single continent. They have been greatly modified by later evolution, however, so they do not really qualify as living fossils.

The Mud

Back at Gibraltar, the crew adjusted their instruments and used the newly available telegraph cable to measure the time it took for an electrical signal to reach the Crown Colony of Malta; it corresponded to 1,600 km. *Challenger* ventured only a short distance into the Mediterranean Sea before turning around and heading west into the Atlantic. On January 23, 1873, the vessel had left the shallow waters of the continental shelf and was sailing over the great deep of the Atlantic Ocean. Some 400 km west of Tenerife, the largest of the Canary Islands, at 2,000 fathoms the dredge brought up whitish mud— or "ooze"—containing countless microscopic shells of the foraminifera ("hole-bearer") genus *Globigerina*. The "forams" would turn out to blanket much of the seafloor and later be indispensable to dating the sedimentary layers that cover it.

The recovery of countless foram shells from great depths seemed to put the final quietus on Forbes's azoic theory. But there remained the nagging question of just how life managed to survive in the cold and dark abyss. Wyville Thomson believed that the dredged forams had lived even at several thousand fathoms deep, but there was another explanation: that they had lived at or near the surface and after they died, their calcareous shells had sifted down to accumulate on the seafloor. This was naturalist John Murray's preferred explanation. A dredge hauled up on the way from the Cape Verde Islands to the then-named St. Paul's Rocks, located near the equator roughly halfway across the Atlantic, answered the question: "Nearly the whole of the carbonate of lime present consisted of the dead shells of surface organisms, and it was estimated that 75 per cent was due to pelagic [near surface] Foraminifera."[14] In the best scientific tradition, Wyville Thomson graciously admitted that he had been wrong: "I now admit that I was in error and I agree with [John Murray] that . . . all the materials of such deposits with the exception . . . of animals which we now know to live at the bottom at all depths . . . are derived from the surface."[15]

Further west, water depth remained at about 2,000 fathoms, but then at 1,770 km from Tenerife, the sounding line reached a remarkable 3,150 fathoms. The dredge now no longer brought up the familiar white foram ooze, but a dark silt that the scientists called "red clay." They noticed that the deeper the water dredged, the less ooze and the more clay. Specifically, as Wyville Thomson wrote, "Wherever the depth increases from about 2,200 to 2,600 fathoms, the modern Chalk formation of the Atlantic [the ooze] and of other oceans passes into clay." The scientists poured weak acid onto a sample of *Globigerina* ooze and found that it dissolved the white shells, leaving only a small percentage of red clay. Thomson recognized that ". . . the 'red clay' is not an additional substance introduced from without . . . but that it is produced by the removal, by some means or other . . . of the carbonate of lime which forms probably about 98 percent of the material of the *Globigerina* ooze."[16] Today we know that this reflects the "carbonate composition depth (CCD)": the level at which seawater becomes sufficiently acidic to dissolve carbonate (containing CO_3). Today, the CCD is rising toward the surface as the ocean acidifies due to increased atmospheric CO_2 from fossil fuel combustion, threatening the extinction of countless organisms with carbonate shells.

By 2,655 m farther west, the seabed had risen to 1,900 fathoms, then still farther west, it sank to 3,000 fathoms. This reflected Maury's Dolphin Ridge, which scientists in the next century would rename the Mid-Atlantic Ridge.

In March 1875, sounding in the Mariana Islands of the Pacific Ocean, *Challenger* found a deep oceanic trench. Down, down went the sounding weight, as though it would never hit bottom. It came to a stop at 4,475 fathoms, more than 8 km. The scientifics named it the Swire Deep after sublieutenant Herbert Swire, but the name was later changed to Challenger Deep.

In 2014, graduate student David Barclay was aboard RV *Falkor* in the Marianas, supervising the lowering of two modern scientific instruments into the Challenger Deep. Suddenly, the enormous pressure caused one of the devices to implode. The other instrument serendipitously recorded the sound waves from its dying partner as they bounced off the bottom. You can hear it for yourself.[17] This gave the most precise measurement yet of the Challenger Deep, coming in at 11 km, compared to Everest's elevation of 8.9 km.

Some 90 years later, the mid-oceanic ridges and the deep-sea trenches would be recognized as among the most important features of Earth's surface, both the result of the movement of tectonic plates.

Wholesale Returns

The voyage of HMS *Challenger* led to many scientific benefits and lessons:

- Support of science for its own sake is a worthy goal of philanthropy and of national treasuries. None of the four objectives of the Royal Society for the voyage had anything to do with money; there was no concern for return on investment. This was science at its most pure.

- Scientists cannot understand how Earth works until they explore the two-thirds of its surface covered by water.

- The ocean realm contains surprises—the Mid-Atlantic Ridge, the Challenger Deep, the carbonate compensation depth, and thousands of previously unknown species—whose study would reap rewards. These discoveries suggested that the oceans' floors held other informative surprises, which they did.

- Study of samples of the dredged seafloor sediments suggested that Earth's climate has changed in the past, giving a baseline for comparison to modern, human-caused climate change. Indeed, in 2020, scientists used forams from the *Challenger* collection to study how climate change has affected their evolution into modern species.[18]

- Wyville Thomson's discovery of the lungfish offered further support to Darwinian evolution. So did the great variety of previously unknown species discovered, as it was hard for people of *Challenger*'s day to understand why God would have created so many creatures, yet hidden them out of sight in the great deeps.

- The discovery that the seafloor is covered with soft sediment rather than hard rock suggested to twentieth-century scientists that a hollow tube dropped overboard could plunge into the sediment and pull up a core of the sedimentary layers. This one fact would lead to many of the revolutionary discoveries we are studying.

To paraphrase Mark Twain: "There is something fascinating about science. One gets such wholesale returns . . . out of such a trifling investment. . . ."

3
Coral Reefs: Rainforests of the Sea

Charles Darwin was the first to propose drilling for a scientific purpose—not in relation to his famous work on natural selection, but to assess his earlier theory on the origin of coral reefs and atolls. This was already an important scientific question at the time HMS *Beagle* set sail in 1831 on its five-year voyage around the world, with 32-year-old Darwin as the ship's naturalist and gentleman companion to Captain Robert FitzRoy.[1] The advance notice of the voyage published in the *Athenaeum* called the study of coral reefs "the most interesting part of the *Beagle*'s survey [affording] many points for investigation of a scientific nature beyond the mere occupation of the surveyor."[2] As the only scientist on board, this charge fell naturally to the young Darwin and he took it up with gusto, observing and studying dozens of coral reefs in the Indian and Pacific Oceans. The *Beagle* surveyed the Keeling Islands, two coral atolls near Sumatra, and this first encounter supported the theory for the origin of coral reefs that had already taken shape in Darwin's mind. As the *Beagle* sailed on across the Pacific, Darwin wrote a draft of what would become the first monograph of a prolific writing career: his 1842 book *The Structure and Distribution of Coral Reefs*.[3] The book provides insight into what we might call Darwin's evolutionary mindset well before the publication of *On the Origin of Species* in 1859. He received the prestigious Copley Medal of the Royal Society in 1864, not for his work on evolution, but for *Coral Reefs* and his study of barnacles.

The Origin of Coral Reefs

Nature articles and television programs have familiarized many of us with the colorful and fascinating world of coral reefs. They are undoubtedly photogenic, and they also support some of the most diverse ecosystems on Earth, leading some to call them the rainforests of the sea. Like actual rainforests, coral reefs are threatened today by a variety of human activities, especially rising sea temperatures and resulting ocean acidification.

Though the South Seas are studded with coral reefs of various sizes and shapes, Darwin saw that he could classify them by type—the first step scientists take in trying to understand the origin of diverse but related objects. Some reefs cling to the edges of lofty volcanic islands; others form the dangerous barrier reefs that partially surround low-lying islands, separated from the land by a lagoon; and last are the atolls, the ring-shaped reefs that enclose a lagoon with no visible land inside.

Darwin thought these three types of reefs formed an evolutionary sequence, using the word in its general sense. He wrote that the atolls of the Maldives, a chain of coral islands southwest of Sri Lanka, form "a perfect series, such as we have here traced, [that] impresses the mind with an idea of actual change." This in turn supported his notion "that the theory of subsidence, with the upward growth of the coral, modified by accidents of probable occurrence, will account for the occasional disseverment of large atolls."[4]

As shown in figure 3.1, Darwin's model begins with a living coral reef attached to a volcanic island just below low tide. Coral can live only near the surface, so if the island were to slowly subside, perhaps due to the enormous weight of the volcano, the fringing coral would die and new coral would build atop this no-longer-living platform. As subsidence continues, Darwin wrote, "Fringing-reefs are thus converted into barrier-reefs; and barrier-reefs, when encircling islands, are thus converted into atolls, the instant the last pinnacle of land sinks beneath the surface of the ocean."[5]

Darwin's theory had important implications beyond the origin of coral reefs. As Gordon Chancellor points out, "All of this [the growth of coral reefs] is the result of the accumulation of the calcareous skeletons of untold billions of simple organisms. To paraphrase Darwin, none of these little creatures has the slightest idea

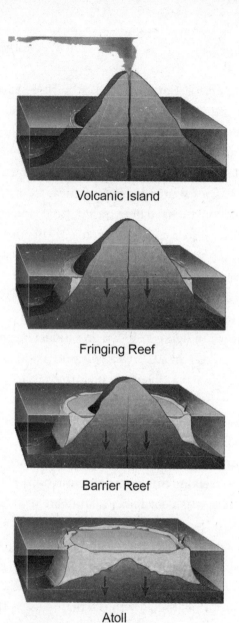

Volcanic Island

Fringing Reef

Barrier Reef

Atoll

Figure 3.1
Evolution of coral reefs and atolls, after Darwin.
Source: Pacific Coastal and Marine Science Center, 2002, United States Geological Survey, https://www.usgs.gov/media/images/atoll-develop ment.

what it is achieving, it just happens."[6] Here is a process of change that takes an enormous amount of time yet has no hand to guide it, only countless creatures doing what comes naturally. Exactly what one could say of natural selection.

The competing theory for the origin of coral reefs at the time came from *Challenger* scientist John Murray. As noted, the expedition had found the oceans rife with microscopic plankton that live near the surface. When these creatures die, their tiny carbonate skeletons rain down continuously onto the seafloor in what would later be called "marine snow." Murray thought that if the deceased plankton sifted down faster than acidic seawater could dissolve their shells, they might build up on submerged rocky platforms and establish a base from which coral could grow. He showed how in his opinion this process could explain the different types of reefs just as well as Darwin's theory.[7]

The debate over the origin of coral reefs became an early example of those knock-down, drag-out controversies that mark the history of science. Behind this particular dispute lay the ulterior motive of opponents of natural selection, who believed that if they could discredit Darwin's coral reef theory, the damage would infect all his scientific work and his reputation.

One who opposed Darwin's theory was Alexander Agassiz, son of the great Harvard zoologist and "Father of Glaciology" Louis Agassiz. The father had become perhaps the most prominent and outspoken scientist to oppose Darwinian evolution and where the father led, the son followed. Alexander Agassiz was himself a scientist but, atypically, had become a millionaire by investing in a Michigan copper mine. In 1876 he visited Murray and learned of his alternative theory for the origin of coral reefs. Agassiz jumped at this opportunity to discredit Darwin, telling Murray, "I never really accepted the theories of Darwin. It was all too mighty simple."[8]

Showing that in those days at least, opposing scientists could still remain on "speaking terms," albeit using pen and ink, in 1881 Darwin wrote to Alexander Agassiz, asking for his judgment on Darwin's theory for the origin of coral reefs:

> If I am wrong, the sooner I am knocked on the head and annihilated so much the better. It still seems to me a marvellous thing that there should not have been much, and long continued, subsidence in the beds of the great oceans. I wish that some doubly rich millionaire would take it into his head to have borings made in some of the Pacific and Indian atolls, and bring home cores for slicing from a depth of 500 or 600 feet. . . .[9]

We might think of it this way: Darwin had not only outlined a theory for the origin of coral reefs but also come up with a way to test the theory and appealed to a wealthy source (he likely had Agassiz in mind) for funding to carry out the project—just what scientists do today when they apply for funding of their research projects.

Darwin believed that if that if his theory was correct, then a core drilled through a coral reef, no matter how deep, would pass through coral until it eventually hit the basalt "basement" rock that forms the upper crust of the ocean basins, below the soft sediment cover. (Basalt is a fine-grained igneous rock composed of the minerals plagioclase, olivine, and pyroxene.) Murray's theory did not envision subsidence, so that the basaltic basement would be hit at much shallower depths.

Agassiz was unwilling to fund a drilling project, but in 1897 the prestigious Royal Society of London commissioned a group of Australian scientists to drill a borehole on the atoll of Funafuti, the capital of the island nation of Tuvalu, whose very existence human-caused global warming and rising sea level now threatens. The Funafuti hole reached 350 m, much of it through "a hard limestone containing numerous well-preserved corals."[10] This seemed to support Darwin's theory, but the drill bottomed out in coral, not reaching the basalt below, so Murray's theory retained a toehold.

Bikini and Eniwetok

Fifty years after the Funafuti drilling, the US government became interested in drilling coral reefs for another purpose: to use them as remote sites at which to test nuclear weapons. In the summer of 1947, to determine the subsurface geological structure before a bomb was detonated, American engineers drilled an 800 m hole on Bikini Atoll in the Marshall Islands, a large archipelago located just north of the equator and west of the International Date Line. The drill passed through limey sediments but again failed to reach basalt. Drilling at another nuclear test site, Eniwetok Atoll, also in the Marshalls, turned up coral all the way down to 1300 m, but this one did reach the basaltic basement. This finally confirmed Darwin's theory of subsidence, though by this time it had come to be generally accepted.

The most important scientific result of drilling in the Marshall Islands had nothing to do with coral reefs or nuclear explosions. Rather it led scientists to wonder just how deep a hole they could drill and what scientific information such a hole could provide. The scientists who wrote the Bikini report recommended drilling to basement in the center of the circular lagoon, saying that the study of the resulting cores would "result in significant contributions in stratigraphy, structural geology, paleontology, petrology, and tectonophysics."[11] The depth to the basalt basement at Bikini was known to be about 2440 m, but the engineers thought that the technology could be extended to reach that deep. The problem was that the depth of the lagoon at Bikini—about 60 m—was too deep to support the drilling rig. The authors of the report did note that the lagoon was "dotted with relatively flat-topped coral knolls," some of which come within a few meters of the surface. They thought that a barge sunk atop one of these submerged mounds could provide a stable platform for drilling.

Harry H. Hess (1906–1969) of Princeton had just described undersea knolls in a paper titled "Drowned Ancient Islands of the Pacific Basin."[12] Hess was a geologist who, during the last years of World War II, served as captain of the USS *Cape Johnson* in the Pacific. He would become one of the founders of plate tectonics and likely the most influential American geologist of his day.

Cape Johnson carried an echo sounder that automatically measured and plotted the depth of the seafloor as the ship passed overhead on its way to wartime landings in the Marianas, the Philippines, and Iwo Jima. The resulting profiles allowed Hess to discover some "160 curious flat-topped peaks" between Hawai'i and the Marianas, at depths from 900 m to 1800 m below sea level. He named them "guyots" after the nineteenth-century geographer, Arnold Guyot, whose name also graced the geosciences building at Princeton.

As the title of his article suggested, Hess believed that guyots are old, much older than the atolls, and are volcanic islands truncated by wave action and then drowned as the ocean floor subsided. Scientists have found them in each of Earth's oceans save the Arctic. Their subsidence not only confirmed Darwin's hypothesis but would later provide critical evidence for plate tectonics.

By the late 1940s, scientists and engineers had established that it was technically feasible to drill thousands of meters into the rocks of the ocean floor, possibly even deeper than the deepest hole in the Marshalls. Hess and other leading geologists knew that the lack of knowledge about the structure and rock types of the oceanic crust was holding back their science. But drilling the deep ocean floor had never been done, and important questions remained: Was it even doable from a vessel pitching in the ocean waves? And should the hole be as deep as possible, or was there some obvious geologic target?

4
Probing the Seafloor

In the 1920s, German science, which led the world prior to World War I, was attempting to restore its reputation from the ravages of the war and to help the nation find a way to pay the devastating reparations that the victors had imposed. The German Scientific Research Aid Council outfitted the ship *Meteor* and sent it on a mission to determine whether gold could be extracted from seawater which, if it worked, could provide the funds needed to repay the German national debt.[1] This idea was the brainchild of German scientist Fritz Haber, who had won the 1918 Nobel Prize in Chemistry for his discovery that ammonia could be synthesized from nitrogen and hydrogen gases, allowing it to be used in large-scale production of fertilizers and explosives. He had also worked on developing the poison gases that were used with devastating effect during the Great War.

From 1925 to 1927, *Meteor* crisscrossed the Atlantic Ocean from the northern tropics at 20°N to 60°S. It turned out that gold could be separated from seawater, but it was present in such minute quantities that the cost would vastly outweigh the monetary benefit. Nevertheless, the voyage was a scientific success. At 300 locations, the crew measured water temperature and depth and collected water samples and marine specimens, much as *Challenger* had done. The expedition's most notable finding was that the Mid-Atlantic Ridge continues south, then rounds the Cape of Good Hope and heads east toward the Indian Ocean (where it would later be called the Southwest Indian Ridge), suggesting that there might be a worldwide chain of mid-oceanic ridges. *Meteor* carried an early sonar detector that sent pulses of sound into the water and recorded the time it took for them to bounce off the ocean floor and return to the ship. Using the speed of sound in water, the crew calculated the depth of the ocean floor. This "echo sounder" could be run continuously, with individual measurements taken at intervals. (Hess aboard *Cape Johnson* would use an improved version.)

The *Meteor* crew collected piston cores from the seafloor, using a glass tube inside a steel one, suspended by a long string of piano wire with a reusable weight attached that drove the tube into the seafloor. When the apparatus was hauled back to the surface, the glass tube was removed so that the cores could be inspected. But at less than a meter long, they were too short to reveal much about recent earth history.

The first to collect longer cores was the Swedish Deep-Sea Expedition of 1947–1948, aboard the Swedish training ship *Albatross*, a beautiful sailing vessel also equipped with a motor.

Figure 4.1
Albatross, 1948.
Source: GBG University Archive, Wikimedia Commons, February 27, 2008, https://commons.wikimedia.org/wiki/File:Albatross1.jpg.

Oceanographer Björe Kullenberg, working with a skilled instrument maker, designed a new corer that, instead of being merely a metal pipe driven by a heavy weight, used a piston to suction the sediment core back up into the barrel, as illustrated in figure 4.2.[2] *Albatross* recovered some 300 cores 20 m long or longer, supplying research material for decades. The expedition also conducted seismic measurements showing that the thickness of the sediment that covered the seafloor increased with distance from the Mid-Atlantic Ridge. This would turn out to provide important evidence for seafloor spreading and plate tectonics.

To understand the problem Kullenberg's piston corer was designed to resolve, imagine yourself with a sledgehammer and a hollow metal pipe a meter long and several centimeters in diameter with which you intend to sample the layers of sediment in a muddy tidal flat. You hammer the pipe down in the mud, then use a gadget to close off the bottom so that the core cannot slip back out as you pull it up, filled with sediment. You raise the pipe and push the core out, exposing the various layers of sediment. The problem you would face is that as you drove the pipe into the mud, and as you extruded it, the layers would compress together, making them harder to distinguish. The pioneers of gravity coring found that such compaction could reduce the length of a core by half, mushing together the various layers and losing information. It was to get around this problem that Kullenberg invented the suctioning piston corer.

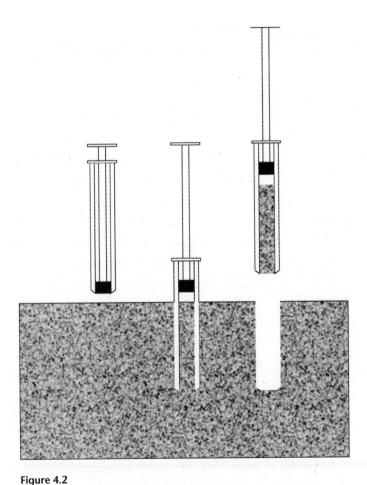

Figure 4.2
Schematic of a piston corer.
Source: Paul J. Somerfield and Richard M. Warwick, "Meiofauna Techniques," in *Methods for the Study of Marine Benthos*, 4th ed., ed. Anastasios Eleftheriou (Hoboken, NJ: Wiley-Blackwell, 2013), 256, https://doi.org/10.1002/9781118542392.ch6.

Kullenberg's key improvement was to install inside the main coring barrel a closely fitted piston attached to a lowering cable. On the left in figure 4.2, the piston is shown near the surface of the sediment. In the center, the piston is held steady while a weight drives the core tube into the sediment. Suction pulls the core up into the tube. The apparatus is withdrawn, and the core laid out on deck. The whole operation takes only a few minutes. An ingenious device, yet so simple, letting gravity do most of the work. One of the most important figures in twentieth-century ocean science, Maurice Ewing, would improve the piston corer even further.

Maurice "Doc" Ewing

Born in 1906 in the small Texas town of Lockney, Maurice "Doc" Ewing was such a mediocre high school student that colleges rejected his applications. His math teacher intervened and convinced Rice Institute (now Rice University), to admit the 16-year-old and provide a scholarship. When he was ready to graduate with a degree in physics and pondering which branch to pursue in graduate school, one of Ewing's teachers urged him to stick to theoretical physics, telling the young man that he had no ability for empirical study. As his memorialist and long-time colleague, Sir Edward Bullard would write, "Rarely has a professor given worse advice," as Ewing would go on to become one of the greatest empiricists of twentieth-century earth science.[3]

Ewing stayed at Rice and earned his PhD in 1931 with a dissertation on seismic waves. He went on to become professor of physics at Lehigh University, where he continued his seismic research by setting off explosives and studying the travel times of the resulting waves through Earth. The first problem that Ewing tackled was to try to use seismic waves at sea to study the subsurface geology beneath the edge of the continental shelf where it slopes down toward the deep ocean floor. Was the slope the result of a geologic fault—a structural feature—or had it formed as sediment gradually spilled out from the eroding continents? That scientists had not answered this question as late as the mid-1930s showed how little they knew about the rocks on and beneath the ocean floor.

Ewing's first attempts to conduct seismic work at sea were unsuccessful. Then the director of the Woods Hole Oceanographic Institution agreed to lend the 45 m steel research vessel *Atlantis*, after its regular season ended. Ewing's research showed that the rock layers continue gently downward under the continental shelf, where 3.7 km of sediment bury them, with no sign of a giant fault.

During World War II, Ewing's work focused on the propagation of sound in the ocean and in 1944, he accepted a position as associate professor of geology at Columbia University. In 1948, a generous donation allowed the university to acquire the Palisades, New York estate of Thomas Lamont just across the Hudson River from the university. A second gift in 1969 from the Henry L. and Grace Doherty Charitable Foundation led to a change of the institution's name to the Lamont-Doherty Geological

Observatory and in 1993 to the Lamont-Doherty Earth Observatory. Doherty had made his fortune with Cities Service Company (later Citgo). Ewing used his Texas roots to maintain good relations with the oil industry, which likely helped bring about the Doherty donation. In his memorial to Ewing, Bullard noted that the name "observatory" was just right for the institution, as Ewing was "primarily interested in finding what was there . . . the emphasis was on data gathering and on its immediate interpretation and not on global theory."[4]

Aboard *Atlantis* in 1947, Ewing set about adapting instruments and methods developed during the war for use in probing the ocean floor. He had learned to set off explosives for seismic work at sea, but those methods required the payload to be lowered to the ocean floor and exploded there. Ewing found that instead the crew could throw a fused barrel filled with explosives over the side of the ship and explode it in the water, then measure the seismic waves reflected from the bottom.

David Ericson, one of the scientists on these voyages, described this harrowing process:

> I remember one used to hold the fuse in one's teeth, sitting on the afterdeck. The charge—a half pound of TNT—was in one hand and a lighter in the other. One wasn't supposed to put the fuse in until the last minute. Then you flung it. This continued at half-hour intervals day and night.
>
> [Ewing] would seize every opportunity to get information. There would be two-day intervals where he was going continuously. If the topography became

surprising the ship would circle and try to outline it.
With him you were either going to sink the ship with
too much explosive or discover something interesting.[5]

Ewing had previously shown little interest in sediment cores but, perhaps realizing their potential, was
persuaded to collect a few on the *Atlantis* voyage. They
would be the first of thousands collected by Lamont
research scientists. Ewing thought that a near-surface
seamount near Bermuda would be a good place to core
without taking up too much ship time. The core from the
seamount was, Ewing would recall, "one of the best of
my life."[6] The top few centimeters were recent sediment
deposits, but below that the sediment dated much further back in geologic time: to the Eocene epoch (see the
geologic timescale, figure 18.1). Many geologists, including Ewing, believed that the seafloor was ancient and
contained a continuous record of earth history. Instead,
this early core showed a large gap in the record and nothing older than about 60 million years.

From Bermuda, *Atlantis* sailed eastward, recording
water depth with an improved echo sounder. At first
the depths remained relatively constant at about 2,900
fathoms—evidence of a flat abyssal plain, as oceanographers call such regions of the ocean floor. A few days
later, the sounder revealed the foothills of the Mid-
Atlantic Ridge and the corer brought up a sample of volcanic basalt. Continued echo sounding showed that the
ridge was a submarine jumble of sharp peaks and valleys
running every which way: a kind of undersea badlands

topography and the exact opposite of the abyssal plain that nineteenth-century scientists had thought floored the oceans.

If one had to choose a single insightful decision for which we should remember Maurice Ewing, it was his edict that no matter the main purpose of a scientific cruise, the crew had to collect at least one piston core each day. This produced vastly more cores than scientists could study, so Ewing established a repository to house them and made it available to all researchers. Ewing told Bullard, "I go on collecting because now I can get the money; in a few years it will not be there anymore, then I shall have the material to keep my people busy for years."[7] By the mid-1950s, the Lamont-Doherty Core Repository had already assembled more than 1,000 cores. Today, the collection includes over 20,000 cores from 11,500 sites in all the major ocean basins, including the Arctic.[8]

Lamont's ships not only collected cores but measured slight differences in the pull of gravity, the heat flowing out of the ocean floor into the water above, undersea topography, and tiny variations in the strength of Earth's magnetic field, all while continuing the original seismic studies. Each would be critical to the development of plate tectonics.

Isotopes

The core collectors hoped to solve one of the most important and long-standing mysteries of science: the

Figure 4.3
Maurice Ewing (foreground) at sea, readying the piston corer.
Source: Woods Hole Oceanographic Institution.

cause of the advance and retreat of huge continental-scale glaciers during the Pleistocene epoch of earth history, 2,580,000 to 11,700 years ago. By studying the microfossils in the cores—some of the same species of foraminifera that the *Challenger* scientifics had collected—modern scientists could learn how the abundance of the forams that thrived in cold water and those that preferred warm water had varied during the ice ages. This gave a general idea of the temperature of the water in which the forams had lived, but a more precise method was needed. Fortunately, at the University of Chicago, Nobel Prize–winning chemist Harold Urey had invented just such a method. It depended on the ratios of oxygen isotopes, which vary with temperature.

The first to apply the oxygen isotope method to ocean cores was an Italian American scientist named Cesare Emiliani, who had been a research associate in Urey's geochemistry laboratory.[9] Emiliani combined his knowledge of subtle differences in the multitude of foraminifera that lived over finite intervals, continually evolving, with his expertise in the new field of oxygen isotope ratios. He used eight cores from the *Albatross* voyage and four from the burgeoning Lamont collection. The oxygen isotope ratios in the forams varied in a zig-zag pattern, reflecting corresponding changes in temperature.[10] The peaks and valleys occurred at roughly the same times as the stages of glacial advance and retreat that geologists had identified. However, the previous studies had recognized only four major advances and retreat of ice age glaciers, while

Emiliani's zig-zag pattern revealed at least seven in the Atlantic and 15 in the Pacific. He conjectured that several processes had combined to cause the glacial cycles: the growth of the continents; the amount of sunlight falling on different regions due to variations in Earth's orientation in space (the Milankovitch theory, to be discussed); positive feedback due to reflective ice replacing sunlight-absorbing water; and the loading of the continents by the heavy ice sheets. As we will see, deep-sea cores would show that one of these factors greatly outweighed the others in importance.

Thus, one of the earliest studies based on deep-sea drilling cores revealed their potential for geological breakthroughs. But by the time Emiliani's findings appeared, leading geologists had diverted their attention toward a more spectacular goal than shallow cores: to drill a hole through Earth's crust to retrieve a sample of the mantle below.

The Mohole

Figure 4.4 shows Earth's surface layers as scientists understood them as of the late 1950s, based on the study of earthquake waves. The broad picture remains accurate today, though the details have changed. The boundary between the crust above and the mantle below is known as the Mohorovičić discontinuity (the "Moho"), after its Croatian discoverer. It marks the level at which

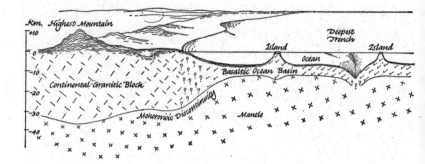

Figure 4.4
Schematic cross-section showing how the Moho lies at a more shallow
depth beneath an ocean basin.
Source: Willard Bascom, *A Hole in the Bottom of the Sea: The Story of the
Mohole Project* (New York: Doubleday, 1961), 23.

earthquake waves speed up by about 15 percent and is
easily detected by seismology. Geologists had a good
idea of the composition of the crust under the continents and ocean basins, but as to the rock type that
makes up the mantle, they could only make educated
guesses. It was an important question, as knowing the
nature of the upper mantle would provide clues as to
how Earth had formed and evolved, whether continents
could have drifted, and so on.

As shown at Eniwetok and other South Pacific sites,
rock drilling (as opposed to piston coring, described
earlier) could reach depths of thousands of meters. Oil
drilling had achieved similarly deep levels from platforms in shallow water and engineers believed that a
hole twice that deep was within the technology's capabilities. Such a hole could penetrate the Moho and

retrieve a sample of the mantle. But where would be the best place to drill it? As figure 4.4 illustrates, the mantle is much nearer the surface under an ocean basin than under a continent, so drilling from a ship at sea seemed the best option.

In 1950, in response to the obvious benefits of scientific research demonstrated during World War II, Congress established the National Science Foundation (NSF). Its mission was "to promote the progress of science, to advance the national health, prosperity, and welfare, and to secure the national defense."[11] This was accomplished by inviting researchers to submit project proposals from which a panel of experts would choose the most promising for funding. In 1957, a group of leading earth scientists gathered to review proposals to the still-new NSF. They included Harry Hess of Princeton and Walter Munk from the Scripps Institute of Oceanography in La Jolla, California, near San Diego. The tedious reviewing process left the two feeling unfulfilled. While most of the proposals were worthy, none promised a breakthrough on the major questions of the earth sciences—including the nature of the upper mantle. Munk suggested a different approach: starting from the top down by asking what project, hang the cost, might lead to such advances. His model for such a project was to drill through the Moho and retrieve a sample of the upper mantle. As others reached up into space and then to the Moon, geologists would drill in the opposite direction to learn fundamental facts about our planet. Such a project would have the advantage of allowing geology to share in the national largesse being directed to "big science" projects such as space flight and atom-smashers.

NSF provided start-up funding under the direction of experienced marine engineer Willard Bascom (1916–2000), who dubbed the enterprise Project Mohole. Test drilling took place from an oil-drilling ship refitted for operation in deep water and named *CUSS I* for the four oil companies that had partnered in its construction: Continental, Union, Shell, and Superior. The tests took place near Guadalupe Island, 240 km west of Baja California. Author John Steinbeck, an ardent amateur oceanographer, came along to record the event for *LIFE* magazine.[12] The myriad fans of his writing will recall Steinbeck's masterful *The Log from the Sea of Cortez*, about another scientific cruise he had joined.

So far, we have mainly discussed piston coring, whereby gravity causes a weighted core barrel to plunge into the soft seafloor sediment and retrieve a core. But the attempt to reach the Moho, and much of the scientific ocean drilling that we will describe in the rest of this book, required drilling through the soft sediment and thousands of meters deep into the hard basement rock below. That necessitated the use of a rotary core drill with a drill bit encrusted with industrial diamonds.

The Moho is nearer the surface under an ocean basin, but that still meant drilling in water thousands of kilometers deep. The most obvious problem that would confront a drillship in such deep water, as one jokester put it, was that the drill string (a series of rigid, connected pipes) would be so long that it would resemble a thin piece of spaghetti dangling from the top of the Empire State Building to the ground. The floating drillship had to be able to stay "on hole"—as close to directly over the drill hole as possible—otherwise the drill string would bend and eventually break at one of its joints, and there would likely be no second chance. Anchors would not work in such deep water, for their chains would suffer from the same problem. Bascom invented a clever dynamic positioning system with four outboard motors mounted at strategic points around the ship's hull. Sensors in the water signaled the motors how to adjust to keep the ship on hole. It worked wonderfully, as *CUSS I* remained for a month within one ship's length above the drill hole. This allowed the crew, operating in 3,560 m of water, to drill several holes, one of which penetrated 183 m through soft sediments and into the basalt below. The job was done on time and within the budget of $1.5 million, a rarity for any big and novel project. The Mohole was off to a great start, with the crew receiving a letter of congratulations from President John F. Kennedy himself.[13]

Steinbeck wrote that he had been so excited as the drilling proceeded that he did not change his clothes for five days—afraid he might miss something. He summed up, "Project Mohole has barely been started, but *CUSS I* shows it can be done." Steinbeck confessed that he had stolen a small piece of the basalt taken from the bottom of the hole, but then had been given another, making the theft both unnecessary and embarrassing. This nondescript chunk of rock, he said, was "more precious to me than any jewel." Steinbeck summed up the future of the Mohole project: "We'll be all right . . . with men like these."[14] Unfortunately, "men like these" would not be the ones to determine the fate of the Mohole.

CUSS I was the high point, or should we say the low point, of Project Mohole. From there, everything went wrong. The project had three sponsors: NSF, the National Academy of Sciences, and the American Miscellaneous Society (AMSOC), a volunteer group of self-selected scientists including Hess and Munk. AMSOC provided oversight but had no funding of its own, nor was it eligible for federal funding.

NSF had set as part of the mission of the project "the drilling of a series of holes in the deep ocean floor, one of which will completely penetrate the earth's crust."[15] To some members of AMSOC and others, this clearly meant that the project should drill several holes, not just one. Bascom later pointed out that he had stressed "in dozens of lectures and articles the importance of exploring the ocean basins by drilling many holes and that the sampling beneath the Moho was a long-range objective."[16] But despite NSF's language, others continued to insist on a single hole to the mantle. The three sponsors could not agree on this fundamental goal, causing a fickle Congress to lose confidence in the Mohole project. Meanwhile, its estimated cost had ballooned and the Bureau of the Budget recommended that "the Foundation withhold further financial commitments . . . until the situation is clarified."[17] Newsweek dubbed it "Project No Hole" and Fortune explained "How NSF Got Lost in Mohole." Congress gave up, and to date no drilling project has come anywhere close to the Moho.

It has become commonplace to regard the Mohole project as a failure. But that is wrongheaded. The only thing actually attempted—the *CUSS I* drilling—succeeded admirably. Then Washington politics and the incompetence and arrogance of some of its backers denied the project any opportunity to follow up its initial success. But the Mohole project did not die in vain.

5
Nothing Beats a Map

By the early 1960s, scientists had been collecting piston cores for nearly two decades, leading to discoveries like Emiliani's determination of the timing of the major Pleistocene glacial boundaries. Research was furthered as the Lamont-Doherty Geological Observatory continued to expand its piston core collection, and other institutions began to collect their own cores. But piston coring only sampled the top few tens of meters of soft sediment. Scientists had inferred that below the sediment lay igneous basalt and indeed *CUSS I* had recovered some, to the delight of John Steinbeck. But questions as to the exact nature of the basalt, whether it was the same everywhere, and its age, were all unanswered.

Continental Drift

As the history of science shows so clearly, a failed experiment or a theory that turns out to be false can still lead to progress. Though Project Mohole failed, the planning had brought together the directors of the leading oceanographic organizations, including Lamont-Doherty, Emiliani's base at the University of Miami, the Scripps Institute of Oceanography, and the Woods Hole Oceanographic Institute. The debate over whether the Mohole project should drill one hole or many had focused attention on the potential of drilling below the sediment cover into the hard rock layer below and at far less cost than attempting to reach the Moho. But the projected cost of an expanded drilling program was still beyond what any one institution could afford. The NSF was the only source of the necessary funding and it was unlikely to choose a single institution to carry out the project. Instead, a collaboration would be necessary. That noble idea could well have foundered on the strong egos involved, but for the sake of science, in 1964 egos were set aside to establish a new organization called the Joint Oceanographic Institutions for Deep Earth Sampling (JOIDES). The initial members were the four institutions above; others would join later.

But what was the most important problem that scientific ocean drilling might solve? The same question could have been asked of the *Challenger* expedition and surely the answer then was whether the biological evidence from the oceanic realm supported Darwin's theory of evolution. The analogous grand question for the earth sciences by the mid-twentieth century was whether continents drift. In 1957, when AMSOC took up the idea of drilling to the Moho, most geologists were committed to uniformitarianism: the theory that the same natural laws that operate today have been consistent throughout earth history and that geological processes such as erosion, deposition, mountain building, and volcanism have always operated at a uniform pace. They rejected continental drift even though the evidence in its favor was mounting and there was no evidence against it. If the science of geology was to progress, it had to establish under which ruling paradigm it would do so: fixed continents or mobile ones?

German scientist and explorer Alfred Wegener introduced the theory of continental drift in a 1912 article and in a book titled *The Origin of Continents and Oceans*, published in 1922.[1] He was influenced by the close fit of the coastlines of Africa and South America, but also assembled a host of evidence from different branches of geology. Just to pick one supporting fact from among many, consider Mesosaurus, an extinct reptile found in nearly identical rocks of early Permian age (see the geological timescale in figure 18.1) in both southeastern South America and southern Africa. Mesosaurus could not have swum the roughly 3,000 km that separate the two continents today, so the most logical explanation was that at some time in the past the two had been connected. But that meant that one or both continents had moved, violating the dogma of uniformitarianism, which eschewed such cataclysmic events. To avoid that, paleontologists conjured up long, snaking isthmuses called land bridges, which they imagined had once connected the widely separated continents but had then

conveniently vanished beneath the waves. This illustrates the lengths to which scientists will sometimes go to
preserve a long-held theory. (For a masterful example of
scientific biography, see Mott Greene's *Alfred Wegener*.[2])

By the early 1960s, new quantitative evidence in favor
of continental drift was quickly accumulating. One of
the most striking examples, shown in figure 5.1, was
a computerized model by Sir Edward Bullard and colleagues at Cambridge University of how well the coastlines on either side of the Atlantic fit together, using the
500-fathom contour.

Mapmakers had long noted that the shorelines of
South America and Africa fit together, but who could say
it was not due to coincidence? The Bullard map dispensed
with this objection. The average misfit amounted to only
about 130 km in an ocean 3,000 km wide (between Brazil
and Liberia), or about 4 percent. When you consider that
the two continents have been apart for scores of millions
of years, a better fit could not have been imagined. In
fact, some geologists regarded it as too good to be true.
The "Bullard Fit" showed at a glance that geologists had
to accept either the possibility of continental drift, or a
huge and improbable coincidence. Nevertheless, many
senior geologists chose the latter.

Bullard was one in a group of scientists who were
instrumental in reawakening interest in continental drift.
He had trained as a nuclear physicist, but, uncertain if he
could find a career in that field in the early 1930s, switched
to geophysics. There, he succeeded magnificently, making

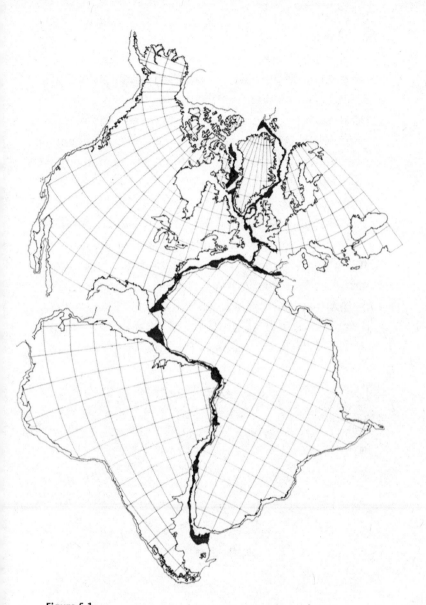

Figure 5.1
The computerized "Bullard Fit." Black = slight gaps or overlaps.
Source: Edward Bullard, James E. Everett, and A. Gilbert Smith, "The Fit of the Continents around the Atlantic," *Philosophical Transactions of the Royal Society of London. Series A, Mathematical and Physical Sciences* 258, no. 1088 (1965): 48.

many scientific advances, as well as founding Churchill College at Cambridge and being knighted in 1953. He and the other physicists whose work helped lead to plate tectonics had the advantage of not having been indoctrinated in the uniformitarian belief that continental drift is impossible.

Another map provided the second half of the cartographic double whammy. The use of the continuous echo sounder at sea to measure the depth to the seafloor had produced enough data to allow two oceanic cartographers at Lamont, Bruce Heezen and Marie Tharp, to construct the first map of the Atlantic floor, shown in figure 5.2. It revealed that a huge, undersea mountain range lay in the center of the Atlantic Ocean basin.

Figure 5.2
The floor of the Atlantic Ocean, 1957, by Bruce Heezen and Marie Tharp. *Credit*: By permission of Marie Tharp Maps, LLC. Fiona Schiano-Yacopino, Nyack, NY.

The Mid-Atlantic Ridge, as it came to be called, appears to have been sliced into a myriad of segments and bent to remain roughly equidistant from the continents on either side. At the very least, the map showed that some gargantuan, catastrophic, nonuniformitarian process had torn the seafloor apart and shoved the pieces around, possibly moving the continents as well.

Marie Tharp, Pioneer

Coming after the Bullard map, the Heezen-Tharp map converted many geologists to continental drift on the spot. The story of how it came to be, and Tharp's role, is worth our attention, but first: Bruce Heezen (1924–1977) had been a junior geology student at the University of Iowa when, in 1947, he met Maurice "Doc" Ewing, who was on a lecture tour (and recruiting campaign for future expeditions) to the hinterlands sponsored by the Sigma Xi scientific society.[3] Heezen had the courage to introduce himself to Ewing, who wasted no time saying, "Young man, would you like to go on an expedition to the Mid-Atlantic Ridge? There are some mountains there, and we don't know which way they run." Heezen leapt at the chance and the following summer found himself chief scientist on a research vessel, despite still being an undergraduate. He returned to Iowa to finish his degree and then joined Ewing's team. Heezen would go on to a distinguished career, only to run afoul of the dictatorial Ewing over some internecine squabble.

Ewing may have had a knack for spotting talent, as he also hired geologist Marie Tharp (1920–2006) to draft maps for Lamont. Tharp has become an important and inspiring figure in the story of women scientists in the twentieth century. Her father was a soil surveyor for the US Department of Agriculture, which required the family to move so often that Tharp attended nearly two dozen schools. In 1943, she graduated from Ohio University with majors in English and music, as well as four minors. The dearth of male students during the war years led the University of Michigan (and others) to admit women, including Tharp, and she received her master's degree in geology in 1944 and went on to work for an oil company. However, she found the work tedious and so left and earned a degree in math from the University of Tulsa. In 1948 she left Oklahoma for New York, where she applied for a job at Columbia University, which, on the basis of her math degree, sent her to interview with Ewing. He must have recognized her talent, but, unsure exactly where to place her with her broad experience, asked, "Can you draft?" She had done some drafting at Michigan, and it was enough for Ewing to hire her. This history may make Tharp seem an impatient sort, but her now-famous maps of seafloor topography speak for themselves—they required extreme diligence and patience for endless corrections of the small mistakes that inevitably creep in when doing finely detailed handwork.

Tharp went to work drafting with Heezen, who had just received his PhD from Columbia. He sailed the ocean gathering data; she remained behind to plot the data from the echo sounder that continually measured the depth to the seafloor on the many voyages of the Lamont ship *Vema*. She plotted the ship's path and noted the depth at each position from the sonar data, giving her a depth profile of the seafloor along the ship's track. She then made a three-dimensional sketch of the seafloor along each profile. Then she assembled these profiles into a map, such as the one shown in figure 5.2.

After Tharp had completed her first six North Atlantic depth profiles (shown in the background in figure 5.3), she saw that each revealed the outline of the Mid-Atlantic Ridge. But she also noticed that each showed a deep cleft right at the center of the ridge. She concluded that this might be a downfaulted rift valley like those in East Africa and might possibly mark the place where the two continents had separated. Ewing was a career-long opponent of continental drift and later plate tectonics and to disagree with him and continue to work at Lamont was not possible. This is the height of irony, or tragedy, since Lamont researchers provided most of the data that would corroborate continental drift and lead to plate tectonics. Heezen had gotten Ewing's message, so that when Tharp showed her putative rift valley to Heezen, he groaned and said, "It cannot be. This looks too much like continental drift." She remembered that he dismissed her interpretation as "girl talk."[4]

Bell Laboratories wanted to lay a new transatlantic telephone cable and asked Heezen to determine the

Figure 5.3
Marie Tharp and Bruce Heezen. Her first six deep-sea depth profiles
are shown in the photo at the right.
Source: Columbia University.

most earthquake-free path. He and his research associate
began to plot up the known locations of North Atlantic
earthquakes, finding that they clustered right in Tharp's
rift valley. This could hardly be a coincidence, instead
showing that the rift was a place of active tectonic
movement. But few were convinced, including famed
oceanographer Jacques Cousteau. He sailed across the
Mid-Atlantic Ridge in his research vessel *Calypso* while
towing a movie camera on a sled along the ocean bottom. What did the film reveal? A rift valley. Cousteau
showed the film at an international ocean congress,
convincing a lot of his fellow oceanographers that the
rift was real.

The irascible Ewing eventually fired Tharp and banned
Heezen, who had tenure, from access to Lamont's research

data. After Ewing moved to the University of Texas, the two returned, producing a map of the Indian Ocean floor and a World Ocean Floor Map. Heezen died of a heart attack aboard ship in 1977, but Tharp continued the work until she died. In 1978, the two were awarded the Hubbard Medal of the National Geographic Society, joining Ernest Shackleton, Louis and Mary Leakey, and Jane Goodall. Marie Tharp's work was instrumental in opening the minds of geologists to the possibility of the forbidden theory of continental drift and to the contribution that women scientists could make. Only in this century has she begun to receive the credit she deserves.

6

Do Seafloors Spread and Continents Drift?

The Bullard and Heezen-Tharp maps changed minds, particularly among younger and more open-minded geologists. But antagonism to continental drift was so entrenched that for the majority to accept it required truly extraordinary evidence. The magnetism of ancient rocks provided that evidence.

During the first half of the twentieth century, scientists had learned that as lavas cool and solidify, certain magnetic minerals within them align their magnetism with that of Earth's magnetic field. Rocks can retain their magnetic alignment for hundreds of millions of years. Think of these minerals as fossil compass needles that point to the North Pole at the time the rock formed. But if North America, for example, had drifted around Earth's surface, then magnetic minerals from the Devonian period (420–360 million years ago), say, would point to the North Pole in a different place from where it lies today. Early studies of ancient rock magnetism showed this was true: the older the rock, the farther from the present North Pole its fossil mineral compass needles pointed. But continental drift was not the only way to explain this observation.

On our human timescale, the magnetic North Pole wanders around the geographic pole, which is the axis of Earth's rotation. Topographic maps display a small arrow showing the direction of the magnetic pole at the time the map was prepared, so that a correction can be made. For all scientists of the 1950s knew, the magnetic pole could have wandered throughout geologic history, explaining why the magnetism of ancient rocks points to somewhere other than the present pole position. Scientists were not even sure why Earth has a magnetic field, so they could hardly rule out the possibility of large-scale "polar wandering." Thus, if the Devonian magnetic pole for North America lay in the central Pacific Ocean, which it did, scientists could not be sure whether it was the continent or the pole that had moved. But if the continents had moved and the pole had not, then Devonian rocks from Eurasia, say, would locate the pole in a different place from the one shown by North American rocks. As illustrated in figure 6.1, this is the case.

The left frame shows the past position of the North Pole with the continents where they are today. The further back in time, the farther the ancient poles for rocks from Eurasia and North America diverge from the present pole, shown as 0°. At any point in time, each continent shows the pole in a different place. The right frame illustrates a further test. If you imagine sliding the continents around until the two polar wandering curves in the left frame

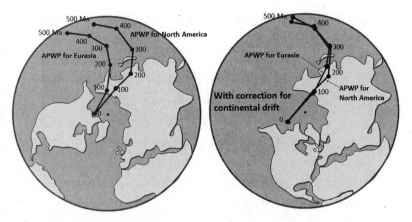

Figure 6.1

Apparent (*left*) and reconstructed (*right*) pole positions. APWP stands for Apparent Polar Wandering Path, that is, one that assumes that the pole and not the continents has moved. Even though the opposite turned out to be the case, this was seen as a convenient (if confusing) way of plotting paleomagnetic data.

Source: Steven Earle, *Physical Geology*, 2nd ed. (Victoria: University of British Columbia, 2019), 329, CC BY 4.0, https://open.bccampus.ca /browse-our-collection/find-open-textbooks/.

lie on top of each other, the continents wind up fitting together into a single protocontinent, which geologists named Pangaea. Just to show how firmly geologists had rejected continental drift, even this indisputable evidence was not enough to convince career-long opponents.

Reversals

To make paleomagnetism more complicated, but also more useful, early in the twentieth century scientists discovered that not all rocks are magnetized so that their mineral compass needles point toward the North Pole; some point in the opposite direction, toward the South Pole. At first, scientists thought this effect derived from "self-reversal"—from some process internal to the minerals themselves that had caused them to switch their polarity. As it became possible to date young volcanic rocks based on the decay of radioactive potassium to argon, scientists discovered that rocks of the same age always have the same polarity: normal (toward the North Pole, like today) or reversed (toward the South Pole). Thus, the opposite polarities meant that rather than the minerals having reversed themselves, Earth's magnetic field has periodically reversed direction. Why this happens is still a mystery. The combination of dating rocks of different ages using radiometric (based on radioactive decay) methods and measuring their magnetic polarity direction led to the paleomagnetic timescale, shown in figure 6.2.

Scientists named the longer subdivisions of the timescale "chrons" (Brunhes, Matuyama, etc.) after pioneers in the study of Earth's magnetism and the shorter subdivisions "subchrons" (Jaramillo, Olduvai, etc.) for geographic locations where the particular reversal was discovered. All rocks from the present back to 0.78 million years ago are normally magnetized: their mineral compasses point toward the present North Pole. Geologists use the paleomagnetic timescale to date rocks based on the pattern of magnetic reversals they show,

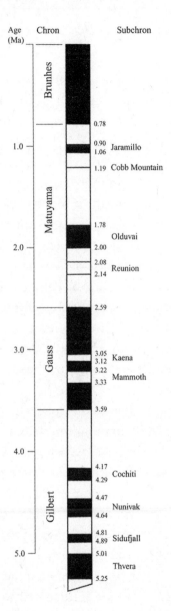

Figure 6.2

The paleomagnetic timescale as of 1979. Black = normal polarity (today), white = reversed.

Source: United States Geological Survey, Wikimedia Commons, October 8, 2007, https://commons.wikimedia.org/wiki/File:Geomagnetic_polar ity_late_Cenozoic.svg.

something like matching two bar codes. Though no one anticipated it, the development of the paleomagnetic timescale turned out to be the clincher for continental drift.

Given the importance of paleomagnetism, vessels on scientific voyages began to tow a magnetometer behind to continuously measure the strength of Earth's magnetic field. That strength combines the intensity of Earth's overall magnetic field and the magnetism of the rocks on the seafloor beneath. If the magnetism in the rocks below points in the same direction as Earth's overall field, the two add and strengthen the field as measured at the surface above that point. This is called a positive magnetic anomaly. If the rocks and the field are oppositely magnetized, they subtract and lower the measured strength: a negative anomaly.

No one had any idea whether magnetic measurements at sea would reveal anything interesting, but since the cost of collecting the data was a minute fraction of the cost of the cruise itself, why not? In 1952, Ron Mason, a professor of geophysics at Imperial College in London, was on sabbatical leave at the California Institute of Technology in Pasadena. While visiting the Scripps Institute of Oceanography, he persuaded its director, Roger Revelle, to have a ship tow a magnetometer behind on an upcoming cruise off the Pacific Northwest. Geologists were interested in this area of the seafloor because it contains huge and mysterious faults hundreds of kilometers long.

Mason and colleagues plotted up their results on a map showing in black where the rocks and the magnetic field lined up to produce a positive anomaly, expecting to see a random, salt and pepper pattern. Instead, they found themselves staring at an odd set of black and white "zebra stripes," as they called them, shown in figure 6.3. The seafloor turned out to be magnetized into broad bands of positive anomalies separated by bands of negative anomalies—far different from the expected salt and pepper. Take a moment to ponder this pattern and try to imagine what could have caused it. Countless geologists did exactly that, at first to no avail.

Mason thought the zones of positive and negative anomalies might represent huge slabs of magnetized rock injected vertically into the seafloor from below, but that seemed *ad hoc* and unsatisfying. He later said he could have kicked himself for not thinking of what in hindsight was the obvious explanation. As often happens, the breakthrough came not from the discoverer of a phenomenon, but from others thinking about it from a different direction. To understand their insight, we need to step back to the early years of continental drift theory.

Seafloor Spreading

As we noted, opponents rejected Wegener's continental drift almost as soon as the theory appeared, not

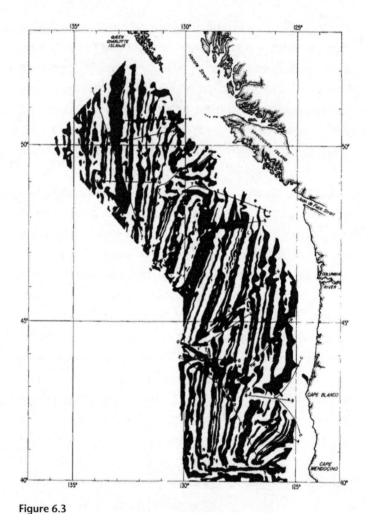

Figure 6.3
The zebra stripes of magnetic intensity. Black shows where the strength of the magnetic field is greater than average.
Source: Arthur D. Raff and Ronald G. Mason, "Magnetic Survey off the West Coast of North America, 40° N. Latitude to 52° N. Latitude," *Geological Society of America Bulletin* 72, no. 8 (1961): 1268.

only because it violated uniformitarianism but also for its "lack of a mechanism"—a known process that could cause continents to move. But, in fact, someone had proposed such a mechanism. In 1928, the great British geologist Arthur Holmes gave a talk to the Geological Society of Glasgow titled "Radioactivity and Earth Movements." Holmes reasoned that since radioactive decay produces heat and since Earth has innumerable radioactive atoms, all inexorably decaying, heat must build up in Earth's mantle and go somewhere, otherwise the planet would warm continuously. "To avoid permanent heating-up," said Holmes, "some process such as continental drift is necessary to make possible the discharge of heat."[1] Thus, Holmes arrived at continental drift not from the fit of the continents or the fossil record, but from physics.

As shown in figure 6.4, Holmes thought the process was convection in the mantle, whereby hot, less dense material at depth rises toward the surface and spreads outward in huge cells that then descend at the margins of the ocean basins. This brings the heat to the surface and allows it to escape into space. The process might rip apart a protocontinent, then the spreading seafloor could move the two resulting continents to either side. But despite Holmes having provided the allegedly missing mechanism, geologists largely ignored his model and in effect convection had to be rediscovered by the next generation.

Let us review the information that geologists had before them in the early 1960s. First, in 1960, Harry

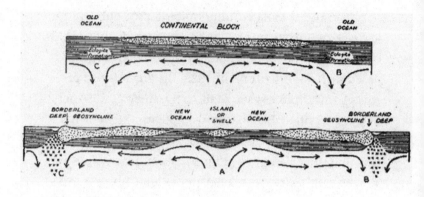

Figure 6.4
Seafloor spreading according to Arthur Holmes.
Source: Arthur Holmes, "XVIII. Radioactivity and Earth Movements,"
Transactions of the Geological Society of Glasgow 18, no. 3 (1931): 579.

Hess postulated, as had Holmes earlier, that the rising
limbs of convecting cells tear a protocontinental block
apart. As the cells move the fragmented parts away
from one another, a median ridge forms as in the Atlantic Ocean, labeled "island or swell" in Holmes's diagram. The continents are swept toward the margins of
the oceans and their leading edges strongly deform as
they descend into the mantle, only to become reheated
and rise again. Second, as shown in figure 6.2, Earth's
magnetic field has reversed its polarity many times over.
Third, the "zebra stripes" reveal that the seafloor comprises alternating bands of greater and lesser magnetic
strength, running roughly parallel to the median ridge.
Geologists everywhere had that information, but the
first to put it all together into a confirmation of seafloor spreading and its corollary, continental drift, were

British geologists Fred Vine and Drummond Matthews. On September 7, 1963, they published an article titled "Magnetic Anomalies over Ocean Ridges." Like many great scientific ideas, their concept was so simple that it took only two sentences to explain:

> If the main crustal layer of the oceanic crust is formed over a convective up-current in the mantle at the center of an oceanic ridge, it will be magnetized in the current direction of the Earth's field. Thus, if spreading of the ocean floor occurs, blocks of alternately normal and reversely magnetized material would drift away from the center of the ridge and parallel to the crest of it.[2]

Thus, each of the zebra bands had been erupted as lava at the crest of a mid-oceanic ridge, solidified and taken on the polarity of the magnetic field at that time, then moved to either side to be replaced by the next eruption, by which time the field had reversed. Figure 6.5 presents a modern diagram of the process in one view.

In the top frame, seafloor spreading has just begun and the volcanic rock near the mid-oceanic ridge is normally magnetized. By the second frame, an episode of reversed magnetism has occurred, followed by a new eruption under normal magnetism, and so on as shown in the third frame. Looking at this from above, one would see the zebra stripes.

In 1964, as JOIDES began its work, seafloor spreading was a new idea and not understood or even noticed by many traditionalist geologists. Maurice Ewing himself, who had more facts and data available to him than

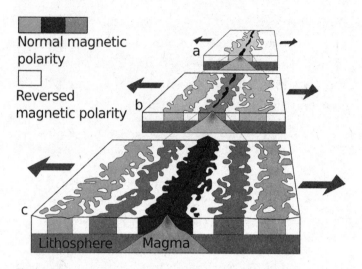

Normal magnetic
polarity

Reversed
magnetic polarity

Figure 6.5
The Vine-Matthews Hypothesis of seafloor spreading.
Source: United States Geological Survey, modified by Chmee2, Wiki-
media Commons, March 2, 2012, https://commons.wikimedia.org
/wiki/Category:Sea-floor_spreading#/media/File:Oceanic.Stripe
.Magnetic.Anomalies.Scheme.svg.

anyone, likely never accepted seafloor spreading, so
one could hardly fault geologists of the time for rejecting or ignoring it. Seafloor spreading was an unproven
hypothesis, a "model" of how things could work, not a
proven fact. But there was a way to test it: if the seafloor
had spread, then the farther from the mid-oceanic ridge,
the older the basement rocks on the seafloor would be.
JOIDES chose testing this hypothesis as the first goal of
the Deep Sea Drilling Project (DSDP).

7
Continental Drift
to Plate Tectonics

The success of *CUSS I* had provided proof of concept, showing that a ship could stay on hole to drill in 3,500 m of water and recover a sample of basalt from the basement rock 180 m below. But the ship's design was outdated, as Steinbeck described when he noted that the ungainly vessel had "the sleek race lines of an outhouse standing on a garbage scow."[1] Moreover, *CUSS I*'s dynamic positioning system, which used suspended buoys as targets, would not work in the much deeper Atlantic. The JOIDES planners knew they needed a modern ship designed specifically for deep-sea drilling. In January 1967, NSF awarded $12.6 million (about $100 million today) for construction of the new vessel, with Scripps as the lead institution.

Scripps contracted with Global Marine Inc. to build the drillship, christened *Glomar Challenger* in honor of the pioneering vessel from 90 years before and shown in figure 7.1. The Levingston Shipbuilding Company laid the keel in Orange, Texas and on March 23, 1968, the new *Challenger* launched and sailed down the Sabine River

Figure 7.1
Glomar Challenger.
Source: Wikimedia Commons, April 27, 2006, https://commons.wiki
media.org/wiki/File:GlomarChallengerBW.JPG.

and into the Gulf of Mexico. After a period of testing, the
leaders of the DSDP accepted the vessel and phase I of
the project was ready to begin. As geologist Edward Winterer put it, "JOIDES advised, Scripps managed as prime
contractor, and NSF monitored—and paid."[2] This three-
headed creature could have spelled trouble, but due to
the breakthrough potential of the program, the three
partners made it work.

To stay on hole in deep water, *Glomar Challenger*
dropped an electronic device called an acoustic transponder (from transmit and respond) to the seafloor. Four

hydrophones near the hull sent signals to the transponder and received signals from it in return. A computer used the arrival times of the signals to adjust the ship's main engine and sideways thrusters to keep it on hole. The system kept the vessel within a circle of 10 percent of water depth even in water several thousand meters deep.

One technological problem that drillers faced in the early days was that the drill bit would wear out—the harder the rock, the faster—and the entire drill string, which could be thousands of meters long, would have to be pulled up and the worn-out bit replaced with a new one. But how to locate the drill hole in deep water so that the string with the new bit attached could be threaded back into it? Clever engineers came up with the idea of installing a reentry cone 5 m in diameter and 4.5 m high into the top of the hole and then to use sonar to locate the cone and reinsert the drill string. Crews tested the method in 1970 in 3,000 m of water off the coast of New York and found it worked wonderfully well.

Baby Steps

Having set out to test seafloor spreading, the JOIDES planners first needed to decide where to drill. By the mid-1960s, scientists had already amassed a large amount of seismic, magnetic, and topographic data about the Atlantic seafloor. Using this information as a guide, scientists planned nine drilling "legs" for the South Atlantic. It would turn out to have a simpler geologic history than other ocean basins, so, in retrospect, the South Atlantic was the ideal place to begin.

The first leg in 1968 was designed to take a few baby steps by exploring a group of buried salt domes in the floor of the Gulf of Mexico, called the Sigsbee Knolls. These domes form when a salt layer under pressure at depth becomes mobile and flows up into a mushroom-shaped structure that arches the sedimentary rock above, sometimes creating reservoirs that trap gas and oil. This first leg showed the capabilities of *Glomar Challenger*. The new dynamic positioning system worked well, allowing the leg to set records for the deepest water yet drilled (5,354 m), the longest drill pipe (5,650 m), and the deepest penetration beneath the seafloor (770 m).[3]

The leg also recovered the oldest rocks yet found in the deep ocean. They were around 150 million years old, placing them in the latest Jurassic, the second period of the Mesozoic era, with the Triassic below and the Cretaceous above. The crew succeeded in locating the salt domes, but the drill soon encountered pockets of natural gas and oil. This raised the frightening possibility that the drill might breach such a reservoir under high pressure, creating a blowout that could destroy the ship and endanger the crew (which is precisely what would happen over 40 years later with the Deepwater Horizon methane blowout in 2010). JOIDES developed a set of procedures to guard against this, but they were no guarantee.

If anyone had thought that deep-sea drilling would be relatively trouble-free, Leg 1 set them straight. The first hole went without difficulty, but then:[4]

Hole 2: "This hole was abandoned and plugged, since there were rigid instructions to avoid any possibility of an uncontrolled flow of oil."

Hole 3: "Abandoned on the basis of an order stating that 2,000 feet (600 m) penetration was the maximum permitted."

Hole 4: "Below about 600 feet (180 m) penetration, the drill with increasing frequency encountered resistant and abrasive chert beds which wore out the bit, and caused operations at this hole to be terminated. The roller bit used here was completely destroyed by the chert." (Chert is a hard rock made of microcrystalline quartz.)

Hole 5: "A diamond bit was used here. The drill bit [got stuck in place] at 260 feet (80 m) of penetration, and drilling and coring were terminated."

Hole 5A: "Once again the chert layers brought drilling to a halt. The cherts and hard limestones were clear signals that deep-sea drilling might involve penetration of very resistant beds."

Hole 6: "Progress was then stopped; the bit was destroyed by radiolarian cherts."

Hole 7: "Drilling and coring operations were terminated because of damage to the core barrel when the vessel was allowed to roll in the trough of a moderate swell."

Hole 7A: "The hole was abandoned on orders from the Captain, who favored a safe margin of travel time to ensure arrival in port at a stated deadline."

And Leg 1 was counted a success!

The leg ended in Hoboken, New Jersey, and Leg 2 departed from there on October 1, 1968, docking at Dakar, Senegal, on November 24. This leg had been designed to provide the critical test of seafloor spreading by determining whether the ages of the basement rocks—below the sediment cover—increased with distance from the Mid-Atlantic Ridge. The seemingly obvious way to do that was to drill down through the deepest (and oldest) sediment layer on the seafloor into the top of the basaltic basement rock below, retrieve a sample of the basalt, and date it by the potassium–argon (K–Ar) method, which works well on young volcanic rocks. It had been the basis of the paleomagnetic timescale shown in figure 6.2, for example. But K–Ar dating requires a special laboratory and could not be done aboard ship. The geophysicists turned instead to their paleontologist shipmates, especially those who were experts on the microfossils that the original *Challenger* had found in abundance. As soon as the core arrived on deck, the paleontologists would retrieve a sample of the sedimentary rock right above the basalt, examine it under a microscope, and identify its characteristic index microfossils, which revealed its age to these experts. That age was used as a proxy for the age of the basalt immediately below. The assumption was that as soon as the volcanic rocks had been extruded at the Mid-Atlantic Ridge and cooled, skeletons of microorganisms would sift down and cover them, so that the bottommost microfossils and the basalt basement would have essentially the same geologic age. As we will see, the results validated this assumption.

Due to mechanical problems that caused the ship to have to return to port for repairs, Leg 2 was not able to drill as many holes as planned. Nevertheless, the drill did reach basalt at three holes, but the scientists were not sure whether it represented true basement or slabs of basalt injected from below, in which case it had not been erupted at the Mid-Atlantic Ridge and could provide no information about seafloor spreading. They did find limey sediments below the present-day calcite compensation depth, where more calcium carbonate dissolves than accumulates. These sediments must have formed at shallower depths and then subsided, as in Darwin's original coral reef theory.

Leg 2 produced a mixed bag, but as it and later legs would show, even a partial success provided new information. And with those lessons learned, Leg 3 hit paydirt.

Paydirt

To decide where to drill to test seafloor spreading, the JOIDES planners used the results of a magnetic survey across the Mid-Atlantic Ridge in the South Atlantic that had been made by scientists aboard the Lamont research vessel *Vema*. These results are shown in figure 7.2. Compare

them with the Vine-Matthews Hypothesis of seafloor
spreading seen in figure 6.5.

The magnetometer produced a continuous record of
the strength of the magnetic field, plotted as "V-18" and
shown on the scale at the right. The curve labeled "V-18
REV" is the same as V-18 except that it has been flipped

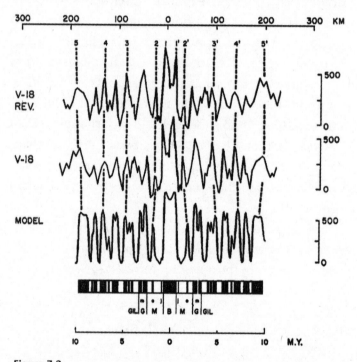

Figure 7.2
Magnetic profile across the Mid-Atlantic Ridge (at 0 km) in the South
Atlantic.
Source: G. O. Dickson, W. C. Pitman III, and J. R. Heirtzler, "Magnetic
Anomalies in the South Atlantic and Ocean Floor Spreading," *Journal
of Geophysical Research (1896–1977)* 73, no. 6 (1968): 2095, https://doi
.org/10.1029/JB073i006p02087.

left to right to show how symmetrical the magnetic profile is on either side of the ridge crest. The "MODEL" profile was calculated assuming a constant rate of spreading and the paleomagnetic timescale shown at the bottom of the chart. It agrees well with the measured profile. The rate of spreading calculated from this information is about 2 cm/year, about as fast as your fingernails grow.

The symmetry of the magnetic profile on either side of the ridge crest was by itself of great significance. Here is one way to put it: if you drilled a hole into the seafloor at some distance west of the ridge crest and then sailed to the other side of the ridge and drilled another hole at the exact same distance east of the crest, the basement rocks from both holes would have the same age, magnetic intensity, and magnetic polarity. Later research would show that even the water depth was close to the same at each site, reflecting uniform subsidence as the seafloor cools, contracts, and becomes denser. These identities could not have come about by chance. Some gargantuan process, centered underneath the ridge crest, had produced a remarkable symmetry out for hundreds of kilometers and scores of millions of years.

By the time *Glomar Challenger* departed Dakar on December 1, 1968, headed for the South Atlantic to commence Leg 3, five years had passed since the Vine-Matthews article had explained the zebra stripes. Some of the modern scientifics aboard accepted seafloor spreading and continental drift, while others did not. In fact, most geologists had hardly heard of seafloor spreading, had been taught that continental drift was false, and were unfamiliar with the new geophysical methods. With the paleomagnetic results from figure 7.2 above as their guide, the *Challenger* crew drilled a series of holes to basement across the Mid-Atlantic Ridge and put seafloor spreading to the test.

One geologist aboard Leg 3 who did not believe in seafloor spreading was sedimentologist Kenneth Hsu. He later wrote a book titled Challenger *at Sea* that gives an insider's perspective on Leg 3 and also on the later Leg 13 in the Mediterranean Sea. In one section, Hsu gives a hole-by-hole account of how the new evidence from Leg 3 forced him against his will to change his mind about seafloor spreading.[5]

The leg's first mid-Atlantic drill site lay 760 km west of the crest of the Mid-Atlantic Ridge. According to the measured rate of spreading of about 2 cm/year from figure 7.2, the seafloor at that site should be about 38 million years old, or latest Eocene. Hsu wrote that he "was secretly hoping it would be much, much older," which would "disprove once and for all the ridiculous theory by the brash young graduate student from Cambridge [Fred Vine]."[6] But when the deepest core came up, its fossils showed the seafloor there to be latest Eocene, as predicted. As tends to happen when someone must confront evidence that shows them to be wrong, Hsu doubled down, attributing the result to coincidence and "await[ing] eagerly" for the next hole to prove it. It lay 420 km from the ridge axis, where the seafloor sediments were predicted to be 21 million years old; earliest Miocene. As the cores at Hole 15 came up from deeper and deeper, Hsu hoped against hope, remembering a saying of the philosopher Karl Popper: "The first 100 swans you see might be white, but should the 101st be black, the theory of white swans is proven wrong." But the age of the seafloor from Hole 15 was also as predicted. As it was for Hole 16 and each of the others. Time after time, figuratively speaking, Hsu would hold his breath, waiting for the next hole to disprove seafloor spreading, and time after time the age result confirmed it. By the time *Challenger* docked at Rio de Janeiro in February 1969, he had become a reluctant convert to seafloor spreading.

The ground was cut from beneath any remaining doubters when the data were plotted as age versus distance from the ridge, as shown in figure 7.3. Except for Hole 21, the farthest from the ridge crest, where the drill was unable to penetrate quite deep enough to reach basement, the ages fall close to a straight line. Most of the holes were located west of the Mid-Atlantic Ridge, but Holes 17 and 18, which fall right on the line, were on the eastern flank. Only one conclusion was possible: the Atlantic seafloor has spread symmetrically from each side of the ridge crest and at a nearly constant rate. This chart would bring about a complete revision in the way geologists thought about the largest features of our planet and what drove them. They understood it at a glance. Prior to the DSDP, almost all the new evidence that pointed toward seafloor spreading had come from geophysics.

Most geologists, trained in other fields, were unsure how to evaluate geophysical evidence, allowing them to preserve their long-held view that seafloors do not spread and continents do not drift. But the chart shown in figure 7.3 was not based on geophysical information. The vertical axis—geologic age—came from something they understood well: the age of the deepest ocean floor sediments, based on familiar microfossils. The horizontal axis was simply the measured distance to the ridge crest. No geologist could claim not to understand this simple chart or, ultimately, to escape its importance.

But that did not make acceptance easy. Hsu was aboard ship, watching the cores come up one by one and hearing the paleontologists tell him over and over that the age of the seafloor at each hole matched that predicted by seafloor spreading. He held the evidence in his hand. If it was this hard for a scientist so close to indisputable evidence to give up a long-held view, imagine how much harder it was for other geologists, far removed from the evidence. As Hsu remembered in his book, his sedimentologist colleagues accused him of being a "traitor to the cause," an "opportunist" who had belatedly jumped aboard "a fashionable bandwagon." But had he refused to "accept the overwhelming evidence," Hsu wrote, he would have been "a conceited bigot who could not admit that he could ever be wrong in his scientific judgement."[7]

But *why* did Hsu and many others initially reject seafloor spreading? Not because they had direct evidence

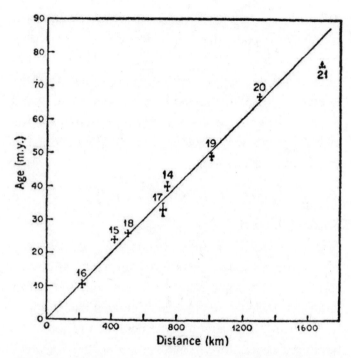

Figure 7.3
Age of basement rock vs. distance from the Mid-Atlantic Ridge, from
Leg 3 of the DSDP. The drill did not reach basement on Hole 21.
Source: A. E. Maxwell et al., "13. Summary and Conclusions," in *Deep
Sea Drilling Project Initial Reports*, vol. III (Washington, DC: US Govern-
ment Printing Office, 1970), 463, http://deepseadrilling.org/03/volume
/dsdp03_13.pdf.

against it, nor because they had come up with another process to explain the "zebra stripes" and other evidence. Hsu said only that he "did not believe the simple formula" of seafloor spreading. They did not *want* to believe in seafloor spreading because to do so would require giving up the dogma of uniformitarianism, the bedrock principle of geology that they had learned at their professors' knees.

Book of Records?

For many geologists, rejection of continental drift went along with the view that the continents and ocean basins are ancient and possibly original, permanent features of Earth. If so, and if neither continents nor ocean basins have ever moved, then seafloor sediments might contain a complete and undisturbed record of earth history all the way back to the beginning of geologic time. Ewing, who knew as much about the geology of ocean basins as anyone, summed up this idea in a 1956 interview:

> In that 2,000 feet of unconsolidated sediment [on the ocean floor] the whole history of the Earth is better preserved than it is in the continental rocks, which have been subjected to heat, folding and mineral changes. . . . The entire record of terrestrial conditions from the beginning of the ocean is there in the most undisturbed form it is possible to find anywhere.[8]

Another geologist remembered:

> In the spring of 1967 when I was a graduate student
> and a year before Leg 1 of DSDP left the dock, one of
> our professors offered to bet anyone in the room $20
> that DSDP would recover a continuous . . . sediment
> section and bottom out in Precambrian basement
> beneath the deep seafloor.[9]

But the scientists aboard Leg 3 did not find a book of
records, only the first few pages. The hole farthest from
the Mid-Atlantic Ridge, site number 21, was the oldest found on the cruise. However, at that site, the drill
did not reach the basaltic basement beneath the seafloor sedimentary rocks, which would have been a bit
older—a good estimate from figure 7.3 would be 80 million years. Thus, the oldest rock found on Leg 3 comes
from the Late Cretaceous, some 14 million years before
the extinction of the dinosaurs. But 80 million years
is only 15 percent of the time to the beginning of the
Paleozoic era and less than 2 percent of the age of Earth
itself. The ocean basins are not old—they are young.
On only its third leg, the DSDP falsified the theory of
permanence and confirmed seafloor spreading. And if
the seafloors move, might they not carry the continents
along, providing the long-sought mechanism of continental drift?

At the time Leg 3 sailed, those who accepted seafloor spreading had no reason not to believe that, like
other geological processes on our mobile Earth, such as
volcanism and earthquake activity, it had sped up and

slowed down over time, perhaps even stopping and starting again. Instead, as shown by the good fit of the data points to a straight line in figure 7.3, spreading in the South Atlantic has gone on for 80 million years at a near-constant rate of about 2 cm/year. This makes two points. First, this rate, calculated using only the ages of the basement rocks from microfossils, is virtually identical to the rate based on the paleomagnetic timescale as shown in figure 7.2. This validates the dating method, the paleomagnetic timescale, and seafloor spreading itself—all three. Second, the constancy of the spreading rate was almost as great a surprise as the fact of seafloor spreading itself. It must mean that the spreading seafloor derives from processes that are among the mightiest acting on our planet, to which the paltry continents and ocean floors are just objects to shove around.

Some geologists imagined running the magnetic tape recorder backward. The magnetic stripes on each side as shown in figure 6.5 would then (in the mind's eye) move in concert back to the ridge crest until finally the last pair met and disappeared. Prior to that moment in geologic time, the Atlantic Ocean did not exist. Not only are ocean basins impermanent, on a geologic timescale they are ephemeral. If we could view Earth from space and fast-forward from the beginning of geologic time to the present, perhaps the most dramatic effect we would note would be the opening and closing of ocean basins and the way the process toyed with the gigantic continents. Today, of course, you can see this in a YouTube video.[10]

In 1968, scientists from Lamont showed that not only the Atlantic Ocean but also the Indian and Pacific Oceans had undergone seafloor spreading, each at a different rate. The global pattern of reversals and seafloor spreading was "in good agreement with continental drift, in particular with the history of the breakup of Gondwanaland."[11] (Gondwanaland, a successor to Pangaea, comprised Africa, Antarctica, Arabia, Australia, India, and South America.)

To Plate Tectonics

The 1960s were a fervent time in geology, with new evidence of all sorts fast arriving. Heezen and Tharp had noted that earthquake epicenters are not located at random but concentrate at the rift down the center of the Mid-Atlantic Ridge. Later evidence would show them also clustered at the deep-sea trenches, continental margins, and along great faults, as shown in figure 7.4. Notice how Earth's surface seems to be divided into regions bounded by narrow belts of earthquake foci. Almost all earthquakes occur where these regions meet, rather than within them. These must be the zones where Earth is in active motion, where one region (or, say, "plate") moves against or away from another, as friction is overcome episodically.

The pioneering Canadian geologist J. Tuzo Wilson put it this way in a groundbreaking 1965 article: "[Earthquake] belts divide the surface into several large rigid plates. The plates between the . . . belts are not readily deformed

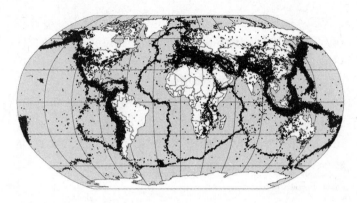

Figure 7.4
Global earthquake epicenters from 1963 to 1998.
Source: NASA, DTAM project team, Wikimedia Commons, December 30, 2004, https://commons.wikimedia.org/wiki/File:Quake_epicenters_1963-98.png.

except at their edges."[12] These statements embody the key insight that turned seafloor spreading and continental drift into plate tectonics. Plates can separate, collide, descend underneath one another, or slide past each other. These motions can account for most of the features we see at the surface, including the different types of mountain ranges, deep-sea trenches, mid-oceanic ridges, and great faults like the San Andreas in California.

Figure 7.5 shows the principal tectonic plates as they are recognized today. Some plates, such as the North American, include both continent and ocean basin, whereas others, such as the Pacific Plate, contain only ocean basin. Thus, it is the plates that move; continents do not move independently of the plates that contain them.

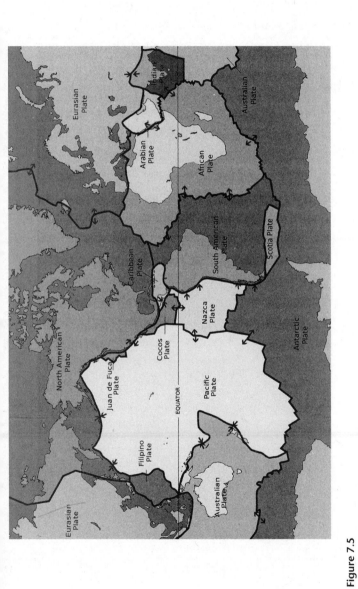

Figure 7.5

Map of the principal tectonic plates.

Source: United States Geological Survey, Wikimedia Commons, February 1996, https://commons.wikimedia.org/wiki
/File:Plates_tect2_en.svg.

The paleomagnetic measurements showed that the ocean basins are rarely older than about 170 million years, just 4 percent of the age of Earth. Those in the Atlantic are seldom older than about 125 million years. This poses a question that could not have been asked before scientific ocean drilling and the advent of plate tectonics: before about 125 million years ago, what existed where we now find the Atlantic Ocean basin? And what happened to it?

8
Hotspots in the Mantle

One year before Vine and Matthews had explained that the zebra stripes were due to seafloor spreading combined with the alternating magnetic field, a philosopher and historian of science named Thomas Kuhn published a book titled *The Structure of Scientific Revolutions*.[1] The book itself caused a revolution in the way historians and philosophers of science comprehend how major changes in scientific understanding occur: not gradually, but more often than not, all at once. His phrase "paradigm shift" soon became an idiom of English language, applied to any number of fields outside of science. Indeed, some would argue that the term has become tiresome, if not a cliché. But there is no doubt that the switch from seeing the continents and ocean basins as ancient and permanently fixed in place, to seeing them as young and mobile, represented just the sort of paradigm shift that Kuhn had in mind.

A scientific revolution occurs when a new theory (or perhaps when an older one, rejuvenated by new data) explains the evidence better than any before it, solving previously unsolvable problems. One long-standing puzzle of geology was the origin of mountain ranges—after the continents and ocean basins the most conspicuous features of our planet. Geologists had long recognized that the Appalachian Mountains, for example, comprise a thick wedge of sedimentary rock that had been compressed and uplifted—but there was no obvious source for the compression. The arrangement of rock layers in the Appalachians showed that the squeezing had come from the southeast, but in that direction lies nothing but open ocean water, which could not have squeezed anything. Despite decades of research, until the advent of plate tectonics, the origin of the Appalachians and similar folded-and-faulted mountain ranges remained intractable. But the inescapable conclusion from plate tectonics is that before the Atlantic Ocean existed, an ancestral sea lay between two converging plates, each carrying a continent. The plates had collided, squeezing up their marginal sedimentary wedges into a mountain range where they met, creating a new and larger plate. Then mantle convection and seafloor spreading had split this plate apart, leaving the Appalachians on the west side of the new ocean—the Atlantic—and the matching Caledonides in Great Britain and Scandinavia on the east side. Thus, the answer to our question of what was present before the Atlantic Ocean is another ocean. And another before that.

But what of the volcanoes of the Cascades, Andes, and others in the Ring of Fire that circumscribes the Pacific Ocean basin? Plate tectonics answers that they were formed when one plate descends ("subducts") so far beneath another that the edges of the plate melt, and the less-dense magma rises to the surface and erupts. But plate tectonics could not explain everything and, most fundamentally, not the cause of seafloor spreading itself. As shown in figure 6.4, in the late 1920s Arthur Holmes had envisioned giant cells of hot magma rising up (convecting) under the mid-oceanic ridges, melting, and spreading to each side, carrying the continents along, then descending under the bordering continents. This model was decades ahead of its time, but understandably proved too simple. Two rival modifications vied to account for seafloor spreading: (1) "ridge push," in which the mantle ascending at the mid-oceanic ridges pushes the seafloor (which may have a continent attached or not) to either side; or, (2) "slab pull," in which the weight of a cool, descending plate at the edge of an ocean basin pulls the rest down along with it, as a section of a blanket hanging over the edge of a bed might drag the entire blanket onto the floor. Slab pull seems to work for oceans that have bordering subduction zones, but for those without them, like the North American Plate, the dominant mechanism must be ridge push.

Island Chains

As we saw above, a key piece of evidence for plate tectonics is that earthquakes and volcanism do not take place at random, but rather where two plates meet. This accounts for volcanic islands on a mid-oceanic ridge crest, as in Iceland. But some oceanic islands, such as those of the Hawai'ian chain, lie in the middle of a plate thousands of kilometers from any earthquake-prone edge. The Pacific has many such remote archipelagos, some made famous during the battles of World War II. They do not derive from volcanism on a ridge crest, so where did they come from?

American geologists believed that James Dwight Dana had long ago supplied the answer. As a member of the US Exploring Expedition of 1838–1842, led by Navy Lieutenant Charles Wilkes, Dana, then only in his late twenties, had visited Vesuvius, the destroyer of Pompeii, and many other islands in the Atlantic and Pacific. The Wilkes expedition spent six months in the Hawai'ian Islands, making Dana the first geologist to explore them. The islands run in a chain from Hawai'i at the southeast end, successively northwest through the major islands of Maui, Lāna'i, Moloka'i, O'ahu, and Kaua'i. The ancient Hawai'ians had deduced from the differing degrees of erosion of the islands in the chain that Hawai'i is the youngest and the others are progressively worn down, and therefore older, to the northwest. According to their mythology, these were the successive subterranean dwelling places of the goddess Pele.

To explain the progression, Dana envisioned a large fissure on the ocean floor through which a succession of volcanic "rents" or tears had opened, allowing lava to pour out onto the surface at different times. The resulting volcanoes had gradually gone extinct, starting with Kaua'i and ending in Hawai'i, where active volcanism is still going on and where the youngest vent must be located. (Mauna Loa on the Big Island erupted as the final touches were being put on this book.) Dana saw the same linear pattern on other Pacific Island chains and came to believe that all had a similar origin. His view of the origin of the Hawai'ian Islands persisted until the arrival of the theory of seafloor spreading.[2]

The geologist who deduced the correct origin of the Hawai'ian Islands is a wonderful example of what a mind suddenly opened to a new idea can accomplish. J. Tuzo Wilson, mentioned at the end of chapter 7, was an urbane and charismatic Canadian who, like all North American geologists of his generation, had been taught that continents are fixed in place and have always been where we find them. But Wilson was willing to change his mind and say so. Such conversions can happen suddenly when a final piece of evidence falls into place to explain the previously inexplicable.

As late as March 1960, Wilson showed himself to be a staunch "fixist," writing, "The view that continents have drifted apart is supported to a small degree but not the extent desired by some advocates of continental drift and some students of paleomagnetism."[3] One of Wilson's students remembered an international geological conference in Helsinki in August 1960 at which Wilson and "mobilist" Felix Vening Meinesz, a Dutch proponent of continental drift, each compiled almost identical lists of geological observations—which each said corroborated his opposite point of view.[4]

Only a year later, at the Pacific Science Conference in Hawaiʻi held in late August 1961, Wilson heard the iconoclastic geologist Robert Dietz read a paper titled "Continent and Ocean Basin Evolution by Spreading of the Sea Floor."[5] As Dietz remembered, "[Wilson] immediately found the concept appealing which surprised me as he had recently written about the impossibility of continental drift. I believe I converted him. . . ."[6] In early October 1961, about 18 months after his article rejecting moving continents appeared, Wilson published a long article endorsing the idea that "major convection currents flow in the mantle, as long advocated by Arthur Holmes, F. A. Vening Meinesz and many others," with the implication that the moving cells carry the continents with them.[7] Wilson did not directly admit to having been wrong, nor testify to his conversion—scientists tend not to do that. He simply adopted convection currents and seafloor spreading as his new and better working model, wrote an article based on them, and gave credit to its proponents. Scientists know that the road to new theories is paved with ones that failed. Thus, there is no need to apologize for having been wrong. Harry Hess exemplified the proper spirit when he wrote, "I take pride in my errors."[8]

His mind now open to new possibilities, Wilson quickly grasped the implications of seafloor spreading. One was that the ocean floor is not the ancient feature that Ewing and generations of geologists had long supposed. Just the opposite: it is geologically young. But how could one test this idea before scientific ocean drilling had brought up cores of the ocean floor? In a 1963 article, Wilson plotted a frequency diagram of the ages of the oldest volcanic islands.[9] It showed that the vast majority are no older than 100 million years, corroborating the youth of ocean basins. Wilson next accepted the hypothesis that volcanic islands had been born on the mid-oceanic ridges and been successively swept to either side, in which case those farthest from the ridge would be the oldest, as DSDP Leg 3 would find for the seafloor itself. A survey of the ages of islands in the Atlantic and Indian Oceans confirmed that this was roughly true, further evidence for seafloor spreading even before Leg 3 had sailed.

Do Hotspots Move?

But many of the linear island chains of the Pacific have no nearby ridges, and for them, Wilson wrote, "No such simple pattern is apparent."[10] The Hawai'ian Islands, for example, lie thousands of kilometers from the East Pacific Rise (the Pacific equivalent of the Mid-Atlantic Ridge). This fact inspired Wilson to yet another new idea that would have been impossible to conceive without seafloor spreading. He noted that there are at least seven linear chains of islands in the Pacific, six of which, including the Hawai'ian chain, "have active or recent volcanoes on the most easterly island and appear to get progressively older towards the northwest in the direction away from the East Pacific rise." Some are shown in figure 8.1.

The two key questions concerned: first, why the islands in each chain line up, and second, why their ages steadily increase from one end to the other, rather than being random. Harkening back to Dana, Wilson

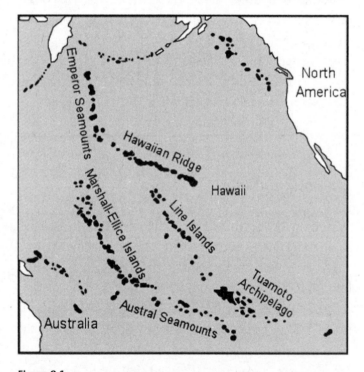

Figure 8.1
Some of the many linear island chains in the Pacific Ocean.
Source: Professor Stephen A. Nelson, Tulane University.

noted that the accepted explanation had been that "each chain of islands had developed by extrusion of lava from a large fault."[11] This putative fissure had gradually opened to the southeast, carrying volcanism with it. But Wilson pointed out that there was no evidence of such fissures beneath the Pacific Islands. Moreover, it was hard to see how a fissure could extend itself to the southeast for tens of millions of years while its northwest end remained stuck in one place.

Dana's hypothesis envisioned the oceanic crust fixed in place while the source of the lavas of the Hawai'ian Islands migrated along the fissure beneath Kaua'i on the northwest to beneath Hawai'i on the southeast. But once Wilson threw off the fetters of fixed continents and ocean basins, he was able to imagine the opposite: the crust had moved while the volcanic source below—the magma-producing hotspot—had remained fixed in the mantle. In a 1964 book chapter (before plate tectonics) Wilson wrote, "The islands are . . . arranged like plumes of smoke . . . carried downwind from their sources."[12] Figure 8.2 illustrates Wilson's model, a testament to how a paradigm shift can open previously unimagined lines of thought.

As shown in figure 8.1, to the northwest, the volcanoes of the Hawai'ian chain link up underwater with the much longer Emperor seamount chain, several of them named for Japanese emperors. The combined Hawai'ian-Emperor seamount chain is 6,200 km long and includes over 80 identified underwater seamounts. Roughly two-thirds

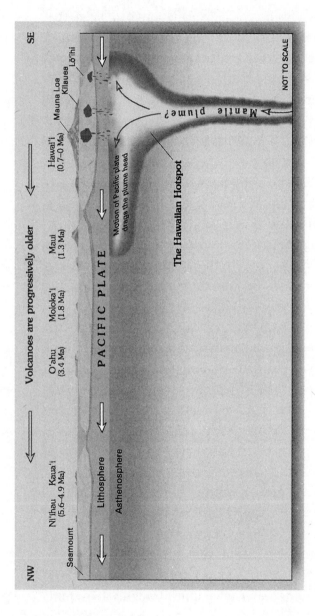

Figure 8.2

Wilson's hotspot model in a modern version. Molten rock rises from the deep hotspot in a fluid plume and intermittently erupts to construct a volcano. The Pacific Plate moves steadily to the northwest so that by the time the next eruption happens, the previous volcano has been moved out of the way and a new one forms in a different place.

Source: United States Geological Survey, Wikimedia Commons, January 1, 2006, https://en.wikipedia.org/wiki/Hawaii_hotspot#/media/File:Hawaii_hotspot_cross-sectional_diagram.jpg.

of the way along the chain, moving northwest from Hawai'i, the chain makes a 60° bend to the north. If Wilson's hotspot model is correct, the bend must reflect a 60° change in the direction of movement of the Pacific Plate over the fixed hotspot below. The ages of the individual Hawai'ian Islands were known by the 1960s, as presented in figure 8.2. But the seamounts were undated. According to Wilson's hypothesis, to the northwest they should get progressively older and more worn down by erosion. Leg 55 of the DSDP collected cores of rock from several members of the Hawai'ian-Emperor chain, dated them, and found that, sure enough, the farther from Hawai'i, the older they were. The data points roughly fit a straight line with a slope corresponding to a constant rate of plate motion of about 8 cm/year. The 60° bend is 43 million years old, but at that time there was no corresponding change in the slope of the curve and, thus, in the speed of the plate. This required that the plate change direction by 60° but maintain the same speed. Was that possible?

In 1971, Jason Morgan of Princeton, who is generally given credit for conceiving of plate tectonics, extended Wilson's hypothesis by showing that a rigid Pacific plate moving over three fixed hotspots could explain three of the Pacific Island chains shown in figure 8.1: the Hawai'ian-Emperor, the Tuamotu-Line Islands, and the Austral-Gilbert-Marshall.[13] This idea appeared to validate Wilson's hotspot model and became a staple of geology textbooks, regarded as evidence for both plate tectonics and hotspots. Its *sine qua non* was that the paternal hotspot for the Hawai'ian chain had remained fixed at the same place in the mantle, below the active volcano Kīlauea on the Big Island. But some of the seamounts are more than 70 million years old. Geologists Peter Molnar and Tanya Atwater wondered how, if the mantle slowly convects, hotspots could remain in the same place for scores of millions of years? Wouldn't the hotspots be dragged along? The two came up with a way to test the idea by comparing the island chains of the Pacific with those of the Atlantic and Indian Oceans.[14]

If a hotspot is fixed in place in the mantle while a plate above moves over it, then at any time in the past, each hotspot should lie under the island chain it created: the Hawai'ian hotspot would have always been under Kīlauea, for example, as implied in figure 8.2. But when Molnar and Atwater used a method of analysis that imagined holding the Hawai'ian hotspot fixed at its location 21 million years ago and figuratively moved other plates to where they were located at that time, the volcanic features of those other plates did not lie over their present hotspots. Instead, the Yellowstone hotspot lay in the Pacific Ocean off the Pacific Northwest, while the Iceland hotspot was closer to Greenland than Iceland, and so on. This is complicated, but the basic conclusion was that hotspots must move with respect to each other and at about the same speed—a few centimeters per year—as the plates themselves. What was needed was an independent method of testing this proposition. Once again, scientific ocean drilling and paleomagnetism came to the rescue.

As we have learned, when lavas freeze, their magnetic minerals line up to point to the location of the North Pole at the time. But magnetism has another indicative feature: its dip or inclination. If you were standing on the equator and held a compass needle that was free to point in any direction—up, down, or sideways—it would point to the pole and also be absolutely level: its inclination would be zero. If you tried this at, say, 45°N latitude, roughly the border between Montana and Wyoming, the needle would point down at an angle of 45°. At the North Pole, it would be vertical. Rock magnetism works the same way: its inclination reveals the latitude at which the rock formed: the "paleolatitude." If the hotspot that created the Hawai'ian-Emperor chain had remained fixed in place while the plate above moved, each island and seamount in the chain, no matter how far from Hawai'i, would have the same paleolatitude: the present-day latitude of Kīlauea: 19.5°N.

Scientists aboard DSDP Leg 55 sampled the 65-million-year-old Suiko Seamount, named for the Japanese Empress who reigned from 593 to 628. Masaru Kono of the Geophysical University of Tokyo made a detailed report on its paleomagnetism, showing that Suiko's paleolatitude of formation was 27°N.[15] This was sufficiently different from Kīlauea's present latitude to corroborate that not only the plate but the Hawai'ian hotspot itself had moved. Leg 145 of the Ocean Drilling Program (ODP), the successor to the DSDP (see appendix for the history of the different scientific ocean drilling programs), focused on the Detroit Seamount, named not for an emperor but for the World War II light cruiser USS *Detroit*. Rocks from the Detroit Seamount dated to 76 million years, making it the second oldest of the Emperor chain, as expected given its far northwestern location. Its paleolatitude turned out to be 36.2°N, even farther from the present latitude of Hawai'i than Suiko.

Leg 197 of the ODP drilled additional holes on the Detroit Seamount, as well as new holes on seamounts Koko and Nintoku, which were younger and closer to Hawai'i. This provided four seamounts whose age and paleolatitudes were known. If the Hawai'ian hotspot had not moved, all four would have had the same paleolatitude as present-day Kīlauea. Instead, each showed a different paleolatitude and the older the seamount, the farther its hotspot had moved.

A leg of the Integrated Ocean Drilling Program in 2010–2011 investigated the Louisville seamount chain in the Southwest Pacific, east of the North Island of New Zealand. It was discovered by sounding only in 1972, but then a radar-bearing satellite located some 70 seamounts in the chain by measuring the altitude of the ocean surface over each. Amazingly, the chain is 4,300 km long. It too has a bend in the middle. The hotspot for the Louisville chain has also moved, but at a slower speed than the Hawai'ian-Emperor chain. Thus, we know that both plates and hotspots move, independently of each other.

Current thinking is that the bends are mostly the result of hotspot motion, with plate motion playing a subordinate role. But it is complicated and views could change. The upside is that in the process of solving such problems, scientists will discover past patterns of circulation in the inaccessible mantle, another result of deep-sea drilling that could not have been found in any other way.[16]

9
When the Mediterranean Dried Up

Ever since maps of the world became available, it was obvious that the facing coastlines of Africa and South America fit together like two pieces of a jigsaw puzzle. This occurred even to Abraham Ortelius, maker of the first world map, who proposed in 1596 that the Americas had been "torn away from Europe and Africa . . . by earthquakes and floods."[1] Many have noticed another tantalizing fact of current world geography: at the Strait of Gibraltar, Africa lies but a few scant kilometers from Europe. It is only natural to wonder whether in geologic history the strait had ever closed and connected the two continents. The Mediterranean Sea depends on the Atlantic Ocean to replace the water the sea loses by evaporation. With the strait closed, could the Mediterranean have dried up? This bizarre possibility occurred to people centuries before Ortelius made his map. In his 37-volume *Natural History*, which would become a model for future encyclopedias, Roman scholar Pliny the Elder (CE 23/24–79) wrote,

> At the narrowest part of the Straits [of Gibraltar]
> stand mountains on either side . . . ; these were the
> limits of the labours of Hercules, and consequently
> the inhabitants call them the Pillars of that deity, and
> believe that he cut the channel through them and
> thereby let in the sea which had hitherto been shut
> out, so altering the face of nature.[2]

Leg 13 of the DSDP, drilling at the sites shown in figure 9.1, would come up with startling evidence that the ancient Iberians had been on the right track.[3]

A Handful of Gravel

Drilling into the bottom of the Mediterranean was an obvious target for an early DSDP voyage. With the new paradigm of plate tectonics as a guide, scientists could use the cores to reinterpret the complex geologic history of the Mediterranean Basin. Another goal was to solve a mystery that had turned up in previous seismic reflection studies: the presence underneath the Mediterranean of salt domes similar to the Sigsbee Knolls that Ewing and crew had drilled in the Gulf of Mexico in Leg 1 of the DSDP. As we noted, these domes well up from a mother bed of salt, forcing their way into the overlying sediments and spreading laterally. European geologists had found candidates for the mother bed in salt formations that outcrop on Sicily and other sites around the margin of the Mediterranean and which might underlie

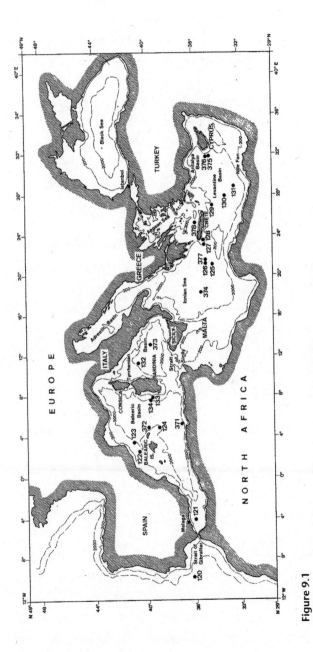

Figure 9.1
DSDP drilling sites in the Mediterranean Sea. Leg 13 sites are numbered 120–134.
Source: K. J. Hsu, Challenger *at Sea: A Ship That Revolutionized Earth Science* (Princeton, NJ: Princeton University Press, 1992), 258.

the sea. Salt is an "evaporite" mineral that forms when bodies of water partially or completely dry up, as in the Dead Sea and Great Salt Lake today. The mother bed in the Mediterranean could have been deposited if an ancestor of the sea had shrunk sufficiently to precipitate salt. The scientists aboard Leg 13 also wanted to solve another puzzle: the nature of a rock layer, nicknamed the "M-reflector," below the floor of the Mediterranean that strongly reflected sound waves, indicating that it was unusually hard.

The cochief scientists of Leg 13 were Kenneth Hsu, whom we met before on Leg 3 as he converted to continental drift, and William Ryan, a 30-year-old from the Lamont-Doherty Geological Laboratory. Ryan was an experienced ocean scientist, having sailed in 1961 on the research vessel *Chain* to explore the Black Sea between Ukraine and Turkey, shown in figure 9.1. The major finding of that cruise was that heavier, saltier water flows at depth from the Aegean Sea through the Bosporus Strait past Istanbul and into the Black Sea, displacing lighter, fresher water which flows near the surface but in the opposite direction. In *Noah's Flood*, written with his longtime research partner Walter Pitman, Ryan describes the Black Sea cruise and provides his own account of Leg 13 to go along with Hsu's.[4]

Glomar Challenger left Lisbon, Portugal on Leg 13 at midnight, August 13, 1970, and returned there less than two months later, on the morning of October 6. The first hole drilled, as shown in figure 9.1 was 121 in the Alboran Sea, a narrow embayment at the extreme west end of the Mediterranean off the Spanish city of Málaga. The second hole was 122, drilled on August 23, 160 km southeast of Barcelona in water 2,300 m deep. The drill penetrated 160 m of soft bottom sediments and the scientists thought it was near the mysterious M-reflector. Sure enough, as Ryan recorded in his logbook, the drill: "Hit something hard!"[5] The core barrel got stuck, and the drill crew had to use a fire hose to jet water into the bottom of the drill pipe to dislodge and retrieve it. The drilling supervisor announced that the hole was unsafe and would have to be abandoned.

Two crew roughnecks unscrewed the bit and washed the mud that had stuck to it into a bucket. With the hole abandoned and nothing to do but wait for breakfast, a sleepless Bill Ryan poured the contents of the bucket into a sink. He gently washed away the mud and examined the gravel-sized fragments that remained. Such seafloor gravels typically accumulate when an undersea avalanche of rock and sediment, called a turbidity current, rushes down a seafloor slope and piles up in a jumble on an abyssal plain below. The gravels typically contain a variety of rock types from the adjacent continent. As Ryan washed and sorted the gravel, he found four types: black oceanic basalt, no surprise since the hole was on the flank of a submarine volcano; white, hardened oceanic ooze, or limestone; elongated crystals that looked like rock candy; and minute fossil shells. None of these could be the result of turbidity currents from the Spanish coast.

Ryan located Italian Maria Cita, one of the leg's paleontologists, who inspected the tiny fossils and told him that they were a "dwarf fauna": mature adults whose growth had been stunted by a stressful environment. This can happen when a high salt content converts seawater into brine: some organisms can still live in it, but they never grow to full size. As Ryan scratched his head, Hsu walked into the lab and immediately noticed the shiny crystals. "Selenite," he declared. Selenite is a crystalline form of gypsum: calcium sulfate with two water molecules. It forms in evaporating brine pools, just the sort of environment that causes dwarfism. But how could brine pools have existed on the floor of the deep Mediterranean? Hsu, Cita, and Ryan next consulted the ship's microfossil specialist, who inspected the dwarf fauna and placed their age as between five and seven million years. This was in the same range as the well-studied evaporite formation exposed on land around the Mediterranean, called the *Gessoso Solifera*. It contains gypsum and rock salt, both of which precipitate when a body of water evaporates past a certain fraction of its original volume. This finding led Hsu to ask his shipmates, "Do you think the entire Mediterranean might once have dried up?"

The English poet William Blake wrote, "To see a world in a grain of sand / And . . . hold infinity in the palm of your hand / and eternity in an hour." Hsu did not go that far, but in a handful of gravel he did see in his mind's eye the entire history of the Mediterranean Sea, the very object of Leg 13. The components of the gravel led Hsu to postulate, as the ancient Iberians had pondered, whether the Mediterranean, blocked from receiving water from the Atlantic, had desiccated and possibly dried up completely, converting the deep-sea calcareous oozes into the limestone that the drill had encountered on its way down. As the sea continued to shrink, it precipitated the selenite and concentrated the brine that had dwarfed the microorganisms. The hard M-layer would then comprise evaporites including gypsum and salt. For this bold and rapid leap based on so little evidence (and turn out to be right), his colleagues nicknamed their shipmate "Instant Hsu."

Any research team is fortunate to include a bold thinker like Hsu. But a group that has the benefit of an "Instant Hsu" also needs a "No Way" Ryan who can say "wait a minute" and call for more evidence. As Hsu tells it, Ryan chided him, "You are too quick to jump to conclusions. There is no need for a theory on the origin of an evaporite, when we have not yet found the evaporite formation!"[6]

The Pillar of Atlantis

After drilling Hole 123 close to 122, *Glomar Challenger* sailed to a point southeast of the Island of Majorca and on August 27 began to drill Hole 124.[7] Drilling proceeded without issue until the bit reached 390 m into the seabed, when again the drill hit something hard enough to shake the entire ship. The pounding of the hard layer went on for six hours, by which time the bit had gained only three more meters. The scientists knew from acoustic sounding that the drill was not deep enough to have hit bedrock. Hsu decided to err on the side of safety and ordered the core barrel withdrawn. The roughnecks expelled the plastic liner and carried the core to the waiting scientists, one of whom exclaimed (referencing another name for the Pillars of Hercules), "By God, we've found the Pillar of Atlantis!"[8] As Hsu remembered, "Lying on the worktable was a beautiful core that indeed resembled a miniature marble column." This was all it took for him to declare, "It is a chicken-wire anhydrite. It says what I said: the Mediterranean dried up then." When Ryan asked for an explanation, Hsu replied, "Oh, chicken-wire anhydrite is a calcium sulfate precipitated by ground water under a sabkha." This only provoked the next obvious question, to which Hsu replied, "Sabkha is an Arabic word for a salt marsh, but they also call their coastal sand flats sabkhas."

Not content with this "chauvinistic" approach (his word), Hsu further flaunted his arcane knowledge by next pointing to a "stromatolite," which few of his listeners had heard of. These, he said, come from algae that secrete huge organic mats over the moist sabkhas after a storm. "Look," Hsu pointed out, "you can even see the outlines of their original cells. They require sunlight for photosynthesis." Here was further evidence that the Mediterranean had once been a series of shallow, briny pools.

The last hole of the cruise, 134, was drilled into the Mediterranean abyssal plain west of Sardinia. It provided the pièce de résistance for the entire leg. As Ryan remembered, "The cores cut from beneath the abyssal plain . . . were the most breathtaking of all, for they contained a deposit that no one, even in a dream, had expected to survive the trip from the seabed to the ship without dissolving."[9]

Indeed, the core did resemble a dream, in that at first the core liner appeared empty. But on closer inspection in better light, it displayed long cylinders of transparent rock. Hsu broke off a piece that looked like an icicle, put it to his lips and tasted, then passed it to his mates to sample. It was pure rock salt that had somehow survived a speedy elevator trip up through 3 km of water. This was the long-awaited mother bed of evaporite salt.

But naturally, not everyone was instantly ready to accept the model that Hsu, Cita, and Ryan had worked out. As the three acknowledged, "The idea that an ocean the size of the Mediterranean could actually dry up and leave a big hole thousands of meters below worldwide sea level seems preposterous indeed."[10] One reason why scientists doubted whether the entire Mediterranean could have dried up and then refilled is the enormous amount of water involved. The abyssal plains of the Mediterranean lie more than 3 km below its surface and the sea holds nearly four million cubic km of water. At first, it is hard to imagine a body of water that size completely evaporating, but then we are not used to thinking in geologic timespans. The Mediterranean loses about 14,000 cubic km of water to evaporation each year. Rainfall and the discharge of the few entering rivers replace about one-tenth of that. The Atlantic Ocean must supply the rest. Had the Strait of Gibraltar closed, which over geologic time would seem eminently possible, if not inevitable in this tectonically active region, simple arithmetic shows that the Mediterranean Sea would dry up in only about 1,000 years. It would leave behind just what the scientists of Leg 13 found: a layer of salt and other evaporites tens of meters thick.

If the entire Mediterranean had dried up, it should have left its mark not just on the floor of the sea but on the streams that entered it. The Mediterranean is the "base level" for those streams. John Wesley Powell, the famed geologist and explorer of the Grand Canyon, introduced this concept in 1875 to describe the lowest level to which a stream can erode. A tributary can erode no lower than the elevation of the larger stream it enters, which therefore is the tributary's base level. When base level falls or the land is uplifted, a river then flows over a steeper slope, gaining erosive power and incising itself deeper into its channel. The Rhône and the Nile are the largest rivers to enter the Mediterranean, which is their base level. If the level of the Mediterranean had fallen rapidly, these ancient rivers would have carved deep canyons, which would have been buried under sediment as the sea refilled. In a search for groundwater near the end of the nineteenth century, drillers discovered a huge gorge buried under the plains of Valence in France, where the Rhône now flows. Ancient erosion had cut the deep canyon into hard granite and subsequently buried it under marine sediments. Scientists eventually traced this buried valley down to the Rhône delta on the Mediterranean at Camargue, where it lay at a depth of 900 m. This perfectly fit the hypothesis of a dried-up Mediterranean. But what of the largest river to enter the Mediterranean, the Nile?

Shortly after *Glomar Challenger* returned to port in October 1970 to end Leg 13, Ryan received a letter from Russian scientist I. V. Chumakov. He had learned of the leg's discoveries from an article in the *New York Times*, portions of which the Russian newspaper *Pravda* had reprinted. Chumakov had been sent to Egypt as part of a Soviet mission to help build the Aswan High Dam, a signature project of Gamal Abdel Nasser, Egypt's ruler after the 1952 coup that he led. In this period of the Cold War, both the US and the USSR courted Egypt and vied to help build the dam, but Nasser thumbed his nose at the US and awarded the project to the Soviets. Searching for rock strong enough to provide footing for the immense foundation the dam would need, the Soviet engineers drilled 15 boreholes across the Nile down into the bedrock below the Nubian Desert. To their shock, to reach hard granite required them to drill through more than 200 m of marine sediment. Where today we find the Nile near Aswan, five million years ago lay a deep gorge, since filled with sediment. At the very bottom of their cores, the Russian scientists had found deep-sea ooze and sharks' teeth. From boreholes drilled in the Nile delta 1,200 km downstream at its mouth into the Mediterranean, the drill went through 3,000 m of sediment without hitting bedrock— deeper than the Grand Canyon. Scientists found similar buried, sediment-filled gorges below rivers in Algeria, Israel, Syria, and other countries that border the Mediterranean. Putting all this together, at the beginning of the Pliocene epoch, the ancestral Nile flowed at a level several thousand meters lower than today. The ancient Mediterranean Sea into which it flowed must have been lower still and possibly dry. How then did it fill to become today's Mediterranean?

Greatest Flood in Earth History?

The concept of an empty Mediterranean basin suggests that the Strait of Gibraltar was once closed. It is not hard to visualize how this could have happened. If the land around Gibraltar were to rise a mere few hundred meters today, the strait would seal and no Atlantic Ocean water could enter the Mediterranean, which would then dry up through evaporation. If later the land fell by a similar amount, the strait would reopen, let in Atlantic Ocean water, and fill the Mediterranean basin. In such a tectonically active region, such uplifts and subsidence would have been the rule, rather than the exception.

Once the strait reopened, how might the Mediterranean have filled? Hsu, Cita, and Ryan thought it had happened quite rapidly. If so, drilling cores would show an almost instantaneous transition from deep-sea sediment oozes above to desert salt and other evaporite minerals below. The core from Hole 134, from which Hsu had tasted salt, brought up a complete section from marine ooze down to basal evaporites. As Ryan wrote, "At that contact the passage from dry desert sand to the marine ooze was razor-thin."[11] Even the bottommost section of ooze had been deposited in abyssal water. Thus, the evidence supported rapid filling rather than gradual. But how rapid? Deep-sea ooze accumulates at a rate of about 2.5 cm per thousand years, causing Ryan to estimate that in no more than 100 years, "The environment changed from dry salt bed to a mile-deep abyss." If he is right, this happened not on a geologic timescale, but in little more than one human lifetime.

The breach that would become the strait might have begun as a trickle of Atlantic Ocean water that found its way across the land bridging the two continents. As the rivulet flowed, stimulated by uplift, it eroded and deepened its channel, allowing more water to flow, causing more erosion and further deepening the channel in a positive feedback loop until the trickle had become a flood that spilled over a giant waterfall and soon filled the Mediterranean Basin. Geologists have recently found evidence for this model in an erosion channel 390 km long stretching from the Gulf of Cádiz on the Atlantic to the Algerian Basin in the western Mediterranean. They estimate the volume of the channel at about 1,000 cubic km and write,

> If the >200 m erosion at the Strait of Gibraltar was caused by the refilling of the Mediterranean Sea, numerical modeling shows that this erosion implies a flood of an unprecedented discharge, possibly above [26 billion gallons per second].[12]

That is 40,000 times the flow over Niagara Falls and 500 times the flow of the Amazon River.

All the material eroded by the Gibraltar waterfall must now reside somewhere on the floor of the Mediterranean. Near the island of Malta, the seafloor drops steeply into a submarine canyon. On the floor of the canyon scientists have found a ~1,600 cubic kilometer accumulation of chaotic sediment that they believe may represent the expected deposits from the megaflood.

We do not have space to do justice to the vast amount of research the drilling on Leg 13 spawned, nor to the many questions that even today remain unanswered. Between 1972 and 2014, a single journal, *Marine Science*, published 153 articles on the great drying of the Mediterranean, which geologists refer to as the Messinian salinity crisis after the Messinian stage of the latest Miocene, itself named for the city of Messina on Sicily. The total number of articles to date for all journals is many times that number.

As any reader of this book will recognize, it is difficult to say which scientific ocean drilling expedition produced the most important results. But those of Leg 13 will find a place on anyone's short list. Some six million years ago, what would become the Strait of Gibraltar closed and the Mediterranean Sea dried up. The evaporating sea water deposited layers of salt several kilometers thick. Then around five million years ago, the strait reopened, allowing Atlantic Ocean water to flood in, refilling the Mediterranean Basin in less than 100 years. There used to be a must-read item in American newspapers called Ripley's Believe It or Not. It presented facts that were true but defied belief, under the motto, "Truth Is Stranger than Fiction."

10
The Glacial Theory

Even before the Deep Sea Drilling Project began in the early 1960s, piston coring (as opposed to rock drilling) by Lamont, Scripps, Woods Hole, and other institutions had already recovered hundreds of shallow cores. These piston cores would be key to solving another of the long-standing mysteries of geology: what caused the ice ages? Before scientists had any reason to ponder that question, they first had to discover that ice thousands of meters thick had once covered much of the Northern Hemisphere. That discovery had its origin in Switzerland, home to deep, U-shaped valleys. Today we know that these landscapes are almost entirely the work of glaciers.

Distant icefields lie at the head of most Swiss valleys. They supply meltwater to small streams that wend their way down through the gravel that lines the valley bottoms. Out beyond the valley mouth lie piles of rock debris of all types, jumbled together in a moraine, from the French for "snout." Early geologists tried to explain these deposits as the result of the great Biblical flood of Noah, whose waters had supposedly receded from the land to fill the ocean basins. One observation hard to explain by "flood geology," however, were the large boulders, sometimes the size of a small house, found on the plains below the valley mouths. Not only would they have been too large for a stream to carry but these exotic stones are often made of a different rock type than the local country rock, sometimes matching types known from far away. Geologists call these migrant stones "erratics," after the Latin for "stray." The greatest geological authority of the nineteenth century, Charles Lyell, came up with a way to make them fit with flood geology. He argued in 1833 that the erratics are "drop-stones" that had been caught in ice floes floating in the global floodwaters, then transported far from their source and released as the ice melted.

In 1831, a student of geology and admirer of Lyell named Charles Darwin accompanied the famous geologist Adam Sedgwick on a field trip to an area of Wales that, as geologists would soon discover, had abundant glacial deposits. The two observers "spent many hours examining all the rocks with extreme care," as Darwin would write in his autobiography, "but neither of us saw a trace of the wonderful geological phenomena all around us."[1] By the time of his next visit to the area, in 1842, Darwin had undergone one of those scientific conversions, now writing, "A house burnt down by fire did not tell its story more plainly than did this valley. If it had still been filled by a glacier, the phenomena would have been less distinct than they now are."

Darwin took from this the important lesson of "how easy it is to overlook phenomena, however conspicuous, before they have been observed by anyone."[2] Even the greatest minds cannot think the unthinkable.

Starting as early as 1787, a succession of scholars, including the German polymath Johann Wolfgang von Goethe, proposed that the contemporary faraway alpine glaciers were the remnants of ancestral ones large enough to fill the valleys and transport rock and sediment from the mountain heights out beyond the valley mouths and drop them below as moraine. To most scientists of the day, this was anathema because it cut out the Great Flood of Noah, rendering the Bible inaccurate and exposing a failure of God's omnipotence.

Nevertheless, at the 1829 meeting of the Swiss Society of Natural Sciences, engineer Ignaz Venetz went even further, arguing that, based on the widespread evidence left behind, the Swiss glaciers had not been restricted to the valleys, but had spread out beyond them to cover large regions of Europe.[3] The attendees roundly rejected his argument, but as sometimes happens in science, one person in the audience, Jean de Charpentier, was immediately persuaded. He kept the seed of continental glaciation alive. Only 15 years before, Charpentier had rejected the hypothesis of widespread glaciation as "not worth examining or even considering." But after listening to Venetz, he was willing to follow the facts and change his mind, the mark of a good scientist. Charpentier assembled the evidence for the glacial theory and presented his conclusions at the 1834 meeting of the society. He told of having been in the process of examining one of the erratics when a local woodcutter happened by. "There are many stones of that kind around here," said the sawman. "They come from far away, from the Grimsel [a high mountain pass in the Bernese Alps] because they consist of Geisberger [granite] and the mountains of this vicinity are not made of it." He told Charpentier of other obvious evidence for larger glaciers in the past, pleasing the scientist so much that he bought the woodcutter a drink with which to toast the distant Grimsel glacier.

Agassiz

The audience at Charpentier's 1834 presentation also rejected the glacial theory, but, yet again, one listener kept an open mind. Louis Agassiz, one of Europe's most famous scientists, had met Charpentier before and admired him. He spent the summer of 1836 at Charpentier's home, studying the fossil fishes which were Agassiz's scientific specialty, now and then examining the glacial evidence that Charpentier showed him. Agassiz not only came to accept the glacial theory but he also became its champion, Charpentier being too retiring to promote his theory. (Though, notably, Agassiz never had such a conversion regarding Darwin's theory of evolution, becoming perhaps the most prominent scientist to oppose it.)

Agassiz quickly assembled the available evidence into a grand theory of glaciation. But like Venetz and Charpentier, he met nothing but obdurate rejection. Agassiz did not help his case by presenting it at a conference of the Swiss society in 1837, having hurriedly written his remarks only the night before. His audience, who had come expecting to listen to Agassiz discussing fossil fishes, instead found themselves hearing of a mind-boggling ice sheet that had covered not only the Swiss Jura but Europe as far south as the Mediterranean. These unexpected and heretical remarks threw the meeting into a tizzy, and on a field trip the next day where, contrary to his expectations, the evidence failed to convince his opponents, Agassiz became so bitter that he walked a quarter of a mile ahead of his distinguished nonfollowers. But after Agassiz's 1837 address, scientists had to confront his glacial theory whether they liked it or not.

In 1840, Agassiz published his monumental *Studies on Glaciers*, laying out the evidence for the glacial theory in detail. But the greatest scientists of the day remained unconvinced. His former professor, the great Alexander von Humboldt, advised Agassiz that he would render a greater service by returning to his work on fossil fishes. "Novel theories," said Humboldt, "convince only those who gave them birth."

British geologist William Buckland joined Lyell in opposing the glacial theory, but when Agassiz met them on their home ground, he soon converted both. He had come to England to attend the 1840 meeting of the British Association for the Advancement of Science. In presenting his paper at the meeting, Agassiz declared that "at a certain epoch all of the North of Europe and also the North of Asia and America were covered by a mass of ice." Lyell sprang to the attack, but Buckland held back. He invited Agassiz and another famous geologist, Roderick Murchison, on a field trip to Scotland and overnight Buckland was converted. Agassiz then showed Lyell unmistakable glacial moraines a mere two miles from the Lyell family estate. Like Darwin, Lyell had been unable to recognize what lay right beneath his nose. Other British geologists would gradually come to accept the glacial theory, as did a new generation of American scientists. The most prominent was Thomas Chrowder Chamberlin, who grew up in Wisconsin and who wrote that he had been "born on a moraine," even closer to the critical evidence than Lyell.

Once the eyes and minds of scientists in Europe and North America had opened to the glacial theory, they saw the evidence everywhere. Some of it suggested that there had been not just one period of glacial advance and retreat but several. For example, in both Britain and America, geologists would often find a layer containing plant remains sandwiched between layers of glacial deposits. This could only mean that a glacier had advanced, then retreated as the climate warmed, allowing soil to develop and plant life to flourish. Then the same glacier or another had advanced again, left deposits, and again retreated. Geologists soon came to recognize

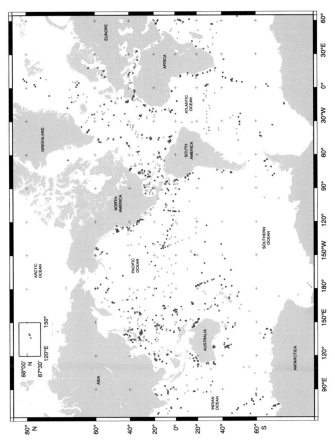

Plate 1

Drilling sites of the three scientific ocean drilling programs.

Source: International Ocean Discovery Program.

DSDP Legs 1–96 (•), ODP Legs 100–210 (•), IODP Expeditions 301–348 (•), IODP Expeditions 349–371 (•)

Plate 2
HMS *Challenger* off the Kerguelen islands in the Southern Indian Ocean.
Hand-colored woodcut.
Source: North Wind Picture Archives / Alamy Stock Photo.

Plate 3
The floor of the Atlantic Ocean, 1957, by Bruce Heezen and Marie Tharp.
Credit: By permission of Marie Tharp Maps, LLC. Fiona Schiano-Yacopino, Nyack, NY.

four periods in which continental glaciers had advanced and retreated. In North America, the most recent is called the Wisconsin glaciation. At its farthest extent in North America, this great ice sheet covered most of Canada, New England, and the Upper Midwest, even reaching the Ohio River. (As described in chapter 4, Cesare Emiliani had found evidence of many more ice advances and retreats.)

The Mystery of the Ice Ages

As scientists came to accept the glacial theory, they naturally asked what could have caused the giant ice sheets. To nineteenth-century scientists, the mystery of the ice ages proved as intriguing as another great enigma of science: what killed the dinosaurs? The cause of dinosaur extinction may hold the record for the number of proffered explanations, one author in 1964 counting 41 and even more would follow.[4] But the various causes put forward to explain the ice ages ran a close second. Indeed, the cause of continental glaciation had more relevance to modern humans, as although the dinosaurs are not about to return, the huge ice sheets will. The discovery that there had been many glaciations, not just one, both complicated and eased the problem. Eased because the repetition itself would turn out to provide a vital clue to the mystery.

The most obvious and perhaps the earliest explanation was that the amount of energy emitted by the Sun had declined, causing global temperatures to fall and glaciers to advance. When the amount of solar energy increased, the glaciers melted and retreated. The discovery of multiple ice ages meant that this waning and waxing of solar energy would have happened repeatedly, but the only solar phenomenon known to repeat is the 11-year sunspot cycle, much too short a timetable to account for the ice ages.

If instead the Sun's output had remained constant, perhaps something had reduced the amount of sunlight reaching Earth's surface. Some scientists thought that increased dust, either in interstellar space or in Earth's atmosphere, might have had such an effect. By the nineteenth century, scientists knew that volcanic eruptions put dust and other substances in the atmosphere, where they absorb the Sun's rays. Nature had even performed an experiment in 1815, when Mount Tambora in the then Dutch East Indies erupted, causing global temperature to fall enough to make 1816 the "Year Without a Summer," leading to food shortages across the Northern Hemisphere. But there was no evident connection between the timing of the ice ages and that of volcanic eruptions.

Or perhaps the culprit was the amount of CO_2 in the atmosphere. By the 1850s scientists knew that CO_2 absorbs heat waves escaping from Earth's surface. A CO_2-rich atmosphere could blanket Earth and raise global temperature, causing glacial ice to melt and retreat. When the amount of CO_2 declined, Earth would cool, and the ice would again advance. In the 1890s, Swedish scientist and Nobelist Svante Arrhenius set out to quantify this effect. Calculating by hand, he found that if the amount of atmospheric CO_2 were cut in half, global temperature would fall by 5–6°C, likely enough to start glaciers on the move. Conversely, were the amount of CO_2 to double, Earth would warm by about the same amount and glaciers would melt. Later, Arrhenius revised the temperature rise from doubled CO_2 downward to 4°C.[5] This "climate sensitivity" is within the range given by today's Intergovernmental Panel on Climate Change.[6] But in Arrhenius's day, no one could think of any reason why the amount of CO_2 in the atmosphere should have repeatedly risen and fallen, so Arrhenius's CO_2 theory failed to gain traction.[7]

These were typical of the difficulties that stood in the way of an acceptable explanation for the ice ages. Some ideas that had once seemed promising had been falsified, others on closer inspection made no sense, and still others were untestable and thus of no scientific value. Then a man self-educated in science took up the problem from a new vantage point. To understand James Croll's contribution, we need to look at Earth as a body in space.

Astronomical Cycles

Over the long run, Earth must release as much heat as it gains from the Sun; otherwise it would become too hot to support life. But in the shorter run, cycles in our planet's orbit and orientation in space cause the amount of solar energy received by the two hemispheres to vary predictably. One such cycle is in the tilt, or *obliquity*, of Earth's axis, which averages 23.44°, as shown in figure 10.1. (The cycles would be named for Milutin Milankovitch, whom we will discuss in the next chapter.) As Earth revolves around the Sun, the axis tilts toward the Sun, causing the Northern Hemisphere to receive more solar energy than the Southern. Six months later, in winter, the axis tilts away from the Sun and the Northern Hemisphere receives less sunlight, while the Southern Hemisphere receives more. The tilt varies between 22.1° and 24.5° on a repeating cycle of about 41,000 years. Without the axial tilt, Earth would have no seasons and temperature would vary only by latitude.

Another cycle occurs because Earth's orbit is not a perfect circle, but rather, as Johannes Kepler discovered in 1609, an ellipse—but only barely. The *eccentricity* of a perfect circle is 0, while that of a straight line is 1. At 0.0167, Earth's orbit is currently almost, but not quite, a perfect circle. But over long periods of time, the gravitational pull of the other planets causes the orbit to become more circular and then less, which affects Earth's nearness to the Sun and the amount of heat received by the two hemispheres. As scientists today know, the cycle of eccentricity has several components: a major one at about 413,000 years and others that combine to create a roughly 100,000-year cycle.

The third cycle comes about because Earth's axis wobbles like the shaft of a spinning top, describing a circle in space. Today the axis points at Polaris, a bright star in Ursa Major. But around 12,000 BCE the axis aimed at Vega, the brightest star in the constellation Lyra. In about the year 14,000, it will again point to Vega.[8] The gravitational pull of the Sun and Moon cause the wobble, known as the *precession* of the equinoxes, or axial precession. Jean le Rond d'Alembert explained in 1749 that axial precession arises because of the pull of the Moon and Sun on Earth's equatorial bulge.[9] He calculated the duration of the axial

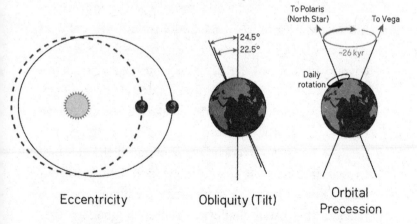

Milankovitch Cycles

Eccentricity   Obliquity (Tilt)   Orbital Precession

Figure 10.1
The Milankovitch cycles.
Source: Skeptical Science, n.d., https://skepticalscience.com/graphics.php?g=342, CC BY Attribution 4.0 International.

precession cycle at 22,000 years, close to the presently accepted figure. The three can be seen in motion in a video from NASA.[10]

Jack of All Trades

The three cycles operate continuously and interact with each other to vary the amount of solar radiation that different regions of Earth receive on a predictable time-table. Before computers, calculating the net effect of the three cycles over time was exceedingly difficult and tedious. One of the first to attempt the calculation was Scotsman James Croll (1821–1890). A self-educated and self-made man, he had worked as a carpenter, millwright, hotelkeeper, tea shop owner, and insurance salesman, at none of which he succeeded, though mostly for reasons beyond his control. Finally, this avid reader of science books landed his ideal position: janitor at the Anderso-nian College and Museum of Glasgow. The post gave Croll access to the museum's fine scientific library and in 1864 he began to focus his reading on the cause of the ice ages. This led him to the work of French astron-omer Urbain Le Verrier, famous for predicting the exis-tence of the planet Neptune. Croll used equations that Le Verrier had developed to calculate changes in eccen-tricity over the last three million years. He found that periods of high eccentricity (less circularity) could last for tens of thousands of years, followed by long periods of low eccentricity. Croll was struck by Earth's present low eccentricity and relatively ice-free condition compared to 100,000 years earlier when both ice-cover and eccentricity were high.

Croll theorized that one hemisphere or the other would undergo an ice age when two conditions were satisfied: (1) high eccentricity, which caused the greatest extremes in the Earth–Sun distance, and (2) a winter solstice that occurred when Earth is farthest from the Sun. He calculated when these two conditions coincided, leading him to conclude that the last ice age had occurred about 80,000 years ago.

Croll doubted whether small changes in Earth's orbit could by themselves affect temperature enough to cause continental glaciation. He hit on the ingenious idea that those small changes could trigger larger ones within Earth's climate system, however, which would feed back and magnify the effect of the orbital changes. (Today we know that climate feedbacks roughly double the effect of atmospheric CO_2 on global temperature, making global warming twice as bad as it would otherwise be.) Croll published his ideas in an 1875 book titled *Climate and Time in Their Geological Relations* and the next year was elected a fellow of the Royal Society of London, the world's most prestigious scientific association.[11] The jack-of-all-trades had found one that he could master and master it he did.

If geologists could come up with a way to date the end of the last glaciation, they could test how close the date came to Croll's prediction of 80,000 years. But at that time, before the discovery of radioactivity, quantitative methods of dating were few and inaccurate. One such crude method depended on knowing how fast waterfalls retreat. They form where a stream runs over a hard rock layer underlain by a softer one. The splashing water erodes the lower layer, undercutting the harder one above, creating an overhang that eventually collapses. This moves the face of the waterfall upstream and the process repeats. From the disposition of moraines, geologists could tell that the Niagara River in New York had cut its famous falls after the last glaciers had receded. Based on the reports of longtime residents of the area, one geologist estimated the rate of retreat of the river at about 1 m/year. Dividing that into the length of the gorge put the time since the last glaciation at about 10,000 years. On his trip to America in 1841, Charles Lyell visited Niagara Falls and revised the estimate to 30,000 years. These estimates were not worth much, but they gave results far lower than had Croll's calculation. With the evidence seeming to run against his theory and with no other way to validate it, scientists essentially lost interest. But the theory sat on the shelf, waiting for just the right person to come along and resurrect it. And come along he did, in the person of a brilliant and indefatigable outsider and trained mathematician. As Arrhenius had done for the "CO_2 theory," he was prepared to devote however much time it took to calculate by hand the effect of the astronomical cycles on global temperature.

11
The Astronomical Pacemaker

Milutin Milankovitch (1879–1958) was a Serbian polymath who made contributions in mathematics, astronomy, climatology, geophysics, and civil engineering. He studied at the Vienna University of Technology, earning his doctorate and writing his PhD thesis on the subject of concrete as a building material. In 1909 he became professor of applied mathematics at the University of Belgrade and, as he remembered, soon found himself "under the spell of infinity and on the lookout for a cosmic problem."[1] He told a colleague that he intended to work out a mathematics-based theory to describe not only the present climate of Earth, Mars, and Venus but those of the past as well. Milankovitch had the mathematical training that Croll and others who had tried to calculate the timing of the astronomical cycles had lacked.

Trip Through the Universe

As he began his calculations in 1911, like Croll, Milankovitch had the benefit of previous computations, in his case made by German mathematician Ludwig Pilgrim. He had calculated the variations in the three critical parameters—obliquity, eccentricity, and precession—back for one million years. With this head start, by 1914 Milankovitch was able to publish four papers on the topic. He showed that the combined effects of eccentricity and precession could produce temperature variations large enough to affect global climate, while axial tilt appeared to play a more minor role.

But before Milankovitch's papers could receive wide circulation, the guns of August blasted forth to launch World War I. While he visited his Serbian hometown, Austro-Hungarian forces captured and imprisoned him. But there are worse places to be during wartime. As Milankovitch remembered, "The little room [his cell] seemed like the nightquarters on my trip through the universe." Then on Christmas Eve 1914 came the welcome news that, thanks to the intervention of a colleague, Milankovitch was to be released and transported to Budapest on condition that he remain there and report in weekly. This gave him four more years of solitude, allowing him to calculate and write "without hurry, carefully planning each step."

In 1920, he published his results in a book titled (translated into English) *Mathematical Theory of Heat Phenomena Produced by Solar Radiation.*[2] Fortunately, it came to the attention of Wladimir Köppen, one of the world's greatest meteorologists and the father-in-law of Alfred Wegener, discoverer of continental drift. The three scientists began a fruitful collaboration in which Köppen persuaded Milankovitch that an ice age occurred when there was too little summer sunlight to melt the previous winter's snowfall.

Milankovitch then set out to calculate the changes in summer radiation for the past 650,000 years. He continued to work and publish and by early April 1941 had a book manuscript in the hands of the printer. The work was to be titled (translated) *Canon of Insolation and the Ice-Age Problem*, insolation being the scientific term for the amount of solar radiation falling on a given area of Earth's surface. Then on April 6, 1941, fate and war again intervened in Milankovitch's life: Germany invaded Yugoslavia, the printing plant was destroyed, and the final pages of his manuscript had to be reprinted.

When his book finally appeared, it clearly showed how the strength of each orbital cycle varied with latitude. It included a chart showing the calculated intensity of solar radiation at three latitudes: 75°, 45°, and 15°. The effect of the 41,000-year tilt (obliquity) cycle is largest at the poles and becomes smaller nearer the equator. The precession cycle has the opposite effect: smaller at the poles and larger at the equator. The 41,000-year cycle dominates the 75° curve while the 22,000-year precession cycle is the major control for the 15° curve. Milankovitch found that the timing of the astronomical cycles from his calculations matched up well with those estimated for the different European ice ages. All this validated his theory, yet it was not enough. Once scientists have made up their minds, as witnessed with seafloor spreading and continental drift, sometimes only irrefutable, quantitative evidence can change them, and for some, not even then. To make matters worse for Milankovitch, early radiocarbon (C-14) dates on glacial deposits appeared to contradict his theory.

Multiple Ice Ages

One of the first measurements using the new radiocarbon dating method, for which Willard Libby won the 1960 Nobel Prize in Chemistry, was made on deposits from the last glaciation in North America. It showed that the ice had reached its southernmost extent about 18,000 years ago, when it covered most of Canada, New England, and the upper tier of Midwestern states. By 10,000 years ago, the ice had retreated far to the north, close to where we find permanent ice today. Milankovitch's curve had predicted that this most recent ice advance had occurred 25,000 years ago. The 7,000-year difference between the radiocarbon age and his calculation could possibly have been due to the slow response time of a colossal ice sheet, like trying to turn an ocean liner 180°. But a layer of peat in Illinois, which could only have been deposited in a warming climate, also gave a radiocarbon age of 25,000 years, contrary to Milankovitch's prediction. As scientists dated older deposits using radiocarbon, more discrepancies arose.

But the radiocarbon method had its own problems. Due to the short half-life of carbon-14, the accuracy of the technique declines with age. Moreover, glacial deposits are inherently difficult to date, as they are a jumble of fragments from the rocks of different ages that the glacier traveled over, scraped up, and mingled together on its long journey, leaving it unclear what in the mix had actually been dated. Scientists who had not given up on the astronomical theory began to search for another way to date these geologically young features. One promising new method employed the radioactive elements uranium and thorium, which could be used to date older samples than could radiocarbon.

Given the difficulty of dating glacial deposits directly, scientists turned to an indirect method. At their farthest advance, the giant continental glaciers locked up an amount of water so immense that the only possible source was the oceans. Thus, the times of greatest ice advance would have been the times of lowest sea level. Conversely, in times of high temperature and glacial melting and retreat, sea level would be at its highest. What was needed was a way to date past high and low stands of sea level, which would correspond to the retreat and advance of continental glaciers. Here we recall Darwin's theory of coral reefs, based on the fact that coral can survive only near sea level. If sea level rises and falls in response to the advance and retreat of continental glaciers, then coral reefs would be repeatedly drowned and exposed. This would leave a series of fossil reef terraces stranded above sea level, each marking a former level of the sea. If Milankovitch was correct, the ages of a set of reef terraces should match those of the peaks in his chart of radiation intensity.

In the 1960s, Wallace Broecker of Columbia University led a team that used the uranium–thorium method to date ancient fossil reefs on Eniwetok, the Florida Keys, and the Bahamas. The results showed that about 120,000 years ago, sea level was some 6 meters higher than today, indicating a time of glacial melting and retreat. Another high level appeared at about 80,000 years ago. The Columbia team joined one from Brown University to determine the ages of fossil reef terraces on Barbados, the easternmost of the Caribbean Islands. One terrace dated to 125,000 years ago and another to 82,000 years ago, close to the earlier results of Broecker's group. Both correspond to periods of high eccentricity and resulting intense solar radiation, and both corroborated the Milankovitch curve. But another Barbados terrace was 105,000 years old, a time not predicted by the Milankovitch solar radiation curve for high latitudes, which the researchers had been using. However, the latitude of Barbados is only 13°N. When the scientists used the radiation curve for lower latitudes, they found a predicted high sea level at 105,000 years ago. Others found terraces with the same three ages on New Guinea (see figure 11.1) and in the Hawai'ian Islands. The different terraces showed episodes of high sea level at about 125,000, 105,000, and 82,000 years ago.[3] This

Figure 11.1
The Huon Reef terraces on Papua New Guinea, a World Heritage Site.
Source: Tezer M. Esat and Yusuke Yokoyama, "Growth Patterns of the Last Ice Age Coral Terraces at Huon Peninsula," *Global and Planetary Change* 54, no. 3–4 (December 2006): 217, https://doi.org/10.1016/j .gloplacha.2006.06.020.

rejuvenated interest in the astronomical theory, but still more evidence would be needed to convince geologists to reconsider a theory they had so long rejected.

Glacial Harmonics

Instead of dating glacial deposits and fossil reefs believed to be due to glaciation and comparing them with the Milankovitch curves, suppose scientists could directly measure past temperature variations. By the mid-1950s, this had become possible using oxygen isotope ratios as a proxy, a method invented in the lab of Nobelist Harold Urey at the University of Chicago. Although all oxygen atoms have 8 protons in their nucleus—that is what makes them oxygen—the element has three isotopes: O-16, O-17, and O-18, with 8, 9, and 10 neutrons respectively. When liquid water evaporates, the lighter O-16 isotope is easier to move and becomes enriched in the vapor. This raises the proportion of O-16 in the atmosphere and in the snow and ice that result when vapor precipitates. The process leaves glacial ice enriched with the lighter O-16 and the ocean enriched with the heavier O-18. The extent of the difference depends on the temperature.

The pioneering work of Cesare Emiliani using oxygen isotope ratios to measure past temperatures, introduced in chapter 4, had shown that piston cores retain a continuous record of historic temperature changes. But his cores went back only 280,000 years. To capture more glacial cycles and truly test the astronomical theory, researchers needed cores that went back further in time. In another development during the fertile 1960s and 1970s, the techniques of paleomagnetism had improved enough to allow them to be used to measure the direction of the weak magnetism recorded in the sediment cores and thus date them using the paleomagnetic timescale shown in figure 6.2. Neil Opdyke, one of the few Lamont scientists open to continental drift, measured

the paleomagnetism of a set of seven piston cores collected by the *Vema* in the deep South Atlantic near Antarctica.[4] In core after core, Opdyke and colleagues detected
each of the major magnetic reversal boundaries and
shorter events back for four million years. The ages of different sections of the cores as deduced from paleomagnetism matched those determined from microfossils. This
was an important milestone, in that lava flows extruded
from volcanoes on land (from where the paleomagnetic
timescale originated), basalts erupted onto the seafloor,
and sediments that had settled through kilometers of
ocean water—all three—recorded the same magnetic
field reversals.

In the 1970s, it was thought that Earth's magnetic
field had last switched direction about 700,000 years
ago, marked by a boundary between the current Brunhes period of normal polarity and the older Matuyama
period of reversed polarity (named for the polarity-
researching geophysicists Bernard Brunhes and Moto-
nori Matuyama, respectively). A core that went back that
far would allow scientists to use the known age of that
boundary as a linchpin from which to calibrate the temperatures represented by the rest of the oxygen isotope
record. Now the wisdom of Ewing's edict of "one core a
day" became apparent. Instead of having to search for a
promising location and drill a deep hole from scratch,
there was a good chance that a piston core for the test
already existed in the rapidly growing Lamont core

collection. One taken from shallow depths in the Western Pacific, labeled V28–238 (V for *Vema*), looked promising. Opdyke located the Brunhes-Matuyama reversal boundary well above the bottom of the core, and James Hays sent samples off to Nicholas Shackleton at Cambridge for oxygen isotope measurement. The result of the collaboration between Opdyke and Shackleton is shown in figure 11.2. In their book, from which this material is drawn, paleoceanographer John Imbrie and his daughter Katherine Palmer Imbrie called this chart "The Rosetta Stone of Late Pleistocene climate."[5]

This result helped to calibrate the ages represented by the variations in oxygen isotope ratios, but could they be used to test the Milankovitch theory? The varying ratios showed some sort of rough pattern of temperature change, but just what was difficult to tease out. The numbers running vertically just to the left of the paleomagnetic timescale are the so-called isotopic stages of the ice ages that geologists have assigned. Between the top of the core and the 700,000-year-old Brunhes-Matuyama reversal boundary, they had identified 19 such stages. This allowed the researchers to interpolate the age of each stage and see how well those ages matched up with the three astronomical cycles of about 100,000, 41,000, and 22,000 years. They were able to identify the 100,000-year cycle in the core, but it so dominated that to find the shorter cycles, a more sophisticated mathematical technique was needed.

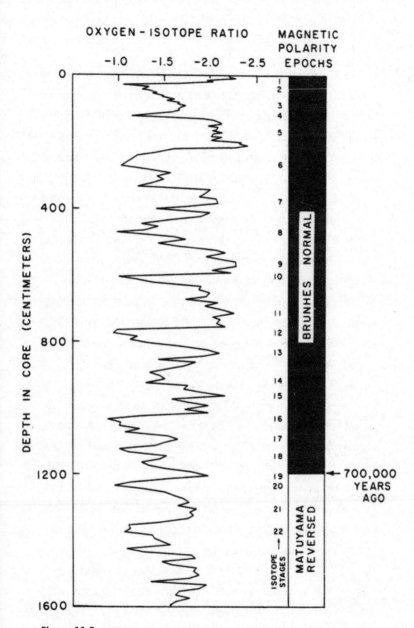

Figure 11.2

Rosetta Stone of Pleistocene climate.

Source: John Imbrie and Katherine Palmer Imbrie, *Ice Ages: Solving the Mystery* (Cambridge, MA: Harvard University Press, 1986), 165.

Dutch scientist E. P. J. van den Heuvel had invented just such a procedure called spectral analysis, analogous to breaking a musical chord into its component notes. Imbrie, now at Brown University, had the necessary computer program and applied it to the "Rosetta Stone" core. He found the dominant 100,000-year cycle peak and also two smaller cycles at 40,000 and 20,000 years ago. This was highly suggestive, but more information would come from a core in which the sediments had accumulated at a faster rate, spreading out the layers of different ages and making the cycles easier to identify. Where could such a core be found? In the Lamont core repository, of course. There, Hays found two cores from the Indian Ocean with the necessary fast sedimentation rates. When Imbrie applied the spectral analysis method, as shown in figure 11.3 he found a 100,000-year cycle from eccentricity; a ~43,000-year cycle from axial tilt; and, it turned out, both 24,000- and 19,000-year cycles, which arise from two different types of precession. (Note that the stated length of these cycles changes slightly with time throughout the book as new methods refine the numbers.) Each of the three dates that emerged from spectral analysis matched predictions of the Milankovitch theory to within 5 percent.

In 1976, Hays, Imbrie, and Shackleton published a paper whose title conveyed their finding: "Variations in the Earth's Orbit: Pacemaker of the Ice Ages."[6] A summary introductory statement read, "For 500,000 years, major climatic changes have followed variations in

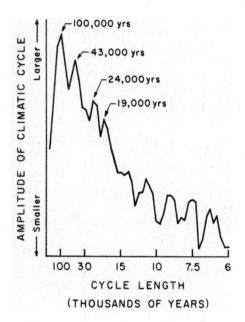

Figure 11.3
Spectral analysis reveals the Milankovitch cycles in a deep-sea core.
Source: John Imbrie and Katherine Palmer Imbrie, *Ice Ages: Solving the Mystery* (Cambridge, MA: Harvard University Press, 1986), 171.

obliquity and precession." At last, scientists had solved the mystery of the ice ages. On the list of the most important scientific papers in geology in the twentieth century, the pacemaker article ranks near the top, right next to the Vine-Matthews paper on seafloor spreading. And scientific ocean coring had provided the critical evidence for both.

12
Astrochronology

Looking at the zigzag plot of oxygen isotope ratios shown in figure 11.2, one could see no evidence of a pattern, just as the untrained ear cannot identify the individual notes within a musical chord. To find the pattern embedded in the isotope ratios required the use of sophisticated, computerized spectral analysis. But the outcome also depended on something more: an undisturbed and complete sediment core. Had erosion removed part of the Rosetta Stone core, or had some burrowing organism disturbed the layering, the spectral analysis would have failed and the Milankovitch theory would not have been confirmed. The hydraulic piston core made the spectral analysis possible. Without it, testing of the Milankovitch theory would have depended on stranded coral reefs and hard-to-date glacial deposits, with results that likely would not have been enough to convince most scientists. In any event, the oldest dated fossil reef was only about 125,000 years old and without the sediment cores there would have been no way to discover whether the orbital cycles influenced climate further back in time. The spectral analysis would lead to an entirely new branch of science: astrochronology.

Tuning

Scientists dated the different levels in the Rosetta Stone core using the paleomagnetic timescale, as shown on the right in figure 11.2. The key date was the Brunhes-Matuyama paleomagnetic boundary at 700,000 years. That age came from potassium–argon dating of basalts in the early 1970s. In this method, the amount of parent potassium and the amount of daughter argon in a specimen are measured. Knowing the rate at which parent changes into daughter, the age of the specimen can be calculated. The method depended on knowing accurately the half-life of K-40, which itself had to be measured, and on the assumption that no daughter argon atoms had escaped the rock samples being dated. Loss of daughter atoms would make a sample appear too young and was a particular problem with argon, a noble gas that rarely forms chemical bonds and whose atoms easily escape a crystal lattice. In other words, there was no way to be sure that the 700,000-year K-Ar age of the Brunhes-Matuyama boundary was accurate. As a check, scientists could date it by another radiometric method, but all methods suffered from the same kinds of problems. But the timetable of the orbital cycles is not measured—it is calculated and must be accurate. Thus, in principle, the results of the radiometric dating methods could be adjusted, or "tuned," to the absolute timetable of the orbital cycles.

Figure 12.1 illustrates one of the first attempts at astronomical tuning. This curve originally looked something like figure 11.2 turned on its side, but mathematical smoothing has removed the extraneous features to leave the dashed line labeled "$\delta^{18}O$." The solid line is the calculated curve of the 41,000-year obliquity (tilt) cycle. Note how the dashed and solid curves show the same pattern but are consistently offset to each other by about 11,000 years. Since the obliquity curve is accurate, the oxygen isotope curve must be tuned to it.

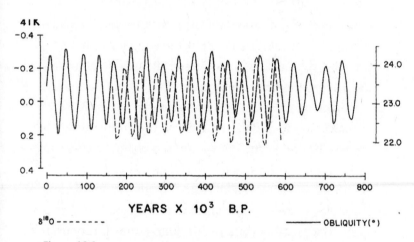

Figure 12.1
Tuning oxygen isotope ratios to the astronomical timescale.
Source: Joseph J. Morley and James D. Hays, "Towards a High-Resolution, Global, Deep-Sea Chronology for the Last 750,000 Years," *Earth and Planetary Science Letters* 53, no. 3 (May 1981): 287, https://doi.org/10.1016/0012-821X(81)90034-0.

Imagine that the dashed oxygen isotope ratio curve is plotted separately on transparent and stretchable paper, so that you could lay it on top and see the obliquity peaks below. You could then slide the oxygen isotope curve over, stretching it as necessary until its peaks coincided as closely as possible to those of the obliquity curve. Then the oxygen isotope ratio curve would have been dated by an astronomical timetable and would be accurate. This would allow it to be compared to other cores that had been similarly tuned to the astronomical timescale. In this composite core, scientists also tuned to the precession cycle (not shown), so that each method served as a check on the other. Each gave the same result.

Once scientists had established the astronomical chronology, they could use it to test the accuracy of the ages that they had assigned to the paleomagnetic boundaries. The youngest is the Brunhes-Matuyama boundary. A new calculation in 1979 had given an age of 730,000 years for the boundary. In 1984, John Imbrie and colleagues, tuning to the precession and obliquity cycles in five deep-sea cores, redetermined the age of the boundary to be 734,000±5,000 years.[1] These varying tuning chronologies were based on short sediment cores that went back to just beyond the Brunhes-Matuyama boundary. A core taken off the coast of Ecuador provided a longer stratigraphic section, back to the Late Pliocene at 2.5 million years. Nicholas Shackleton and colleagues found a particularly strong precession cycle in the core, allowing them to develop a revised timetable that put the Brunhes-Matuyama boundary at 780,000 years and the next major paleomagnetic boundary, the Gauss-Matuyama (see figure 6.2), at 2.6 million years. In all, they dated seven paleomagnetic boundaries using astronomical tuning and found that on average, they were about 7 percent older than the ages given by the potassium–argon method. Shackleton thought this difference was likely due to an inaccurate half-life for the decay of potassium to argon, or possibly to some systemic flaw in the K–Ar method itself.[2]

The K–Ar method requires splitting one rock sample in two, one piece used to measure the amount of potassium, which itself has a limit to its precision, and the other to measure the amount of the daughter isotope argon-40, some unknown amount of which could have escaped. The resulting age combines the inherent inaccuracies in the two measurements, plus a possible imprecision in the half-life of potassium-40, so it is no surprise that it should be systematically off. Searching for an improved method led geologists to a new technique that depended only on the ratios of two argon isotopes and did not require measuring the amount of potassium. But it did require being calibrated against the measured age of a standard and on knowing the half-life of K-40 precisely. In 1992, a team used the improved Ar–Ar method to date the Brunhes-Matuyama boundary at 783,000 ± 11,000 years, close to the age from the astronomical tuning of Shackleton and colleagues.[3] By today, the Ar–Ar method has been steadily improved until ages determined by it match those from the astronomical timescale.

The Astronomical Timescale

By the 1980s, scientists had confirmed the Milankovitch theory and established that it provided the basis for an accurate geologic timescale for the last 50 million years. The Ar–Ar method of dating had replaced the older and less accurate K–Ar method and was being tuned to the orbital cycles. The stage was set to extend the astronomical timescale (ATS) further back in time.

As we saw, in 1990 Shackleton and colleagues expanded the ATS back to 2.8 million years ago, then in 1991 Dutch geologist Frits Hilgen pushed it back to the Pliocene-Miocene boundary at 5.32 million years ago.[4] He studied a group of rocks in Sicily called sapropels, in which dark organic-rich rocks alternate with clayey limestones, as shown in figure 12.2. Note in the figure how the dark sapropels repeat in narrow bands that are themselves bundled into wider sections of light-colored marl, reflecting the effects of the eccentricity and precession cycles. The fine layering of the sapropels enabled Hilgen to date individual layers to within 1,000 years, a much greater resolution than other methods could provide.

To extend and calibrate the ATS back into the Miocene and beyond required complete and undisturbed

Figure 12.2
Milankovitch cycles in sapropels on Sicily.
Source: Alice Marzocchi, "Modelling the Impact of Orbital Forcing on Late Miocene Climate: Implications for the Mediterranean Sea and the Messinian Salinity Crisis" (PhD diss., University of Bristol, 2016), 41.

sections of sediments. With rare exceptions like the Sicilian sapropels, the only source was deep-sea drilling cores. The benefit of having an absolute and accurate timescale was obvious, leading the recovery of such cores to become a principal focus of scientific ocean drilling. In a 2019 article, Kate Littler and colleagues reviewed the results of holes drilled specifically for the purpose of expanding the ATS.[5] These extended the timescale from the Pliocene-Miocene boundary back to the Paleocene about 65 million years ago, with only a few gaps remaining.

The culmination of this research was presented in a 2020 paper by Thomas Westerhold of the University of Bremen and colleagues.[6] Based on cores from 14 scientific ocean drilling sites, Westerhold's group created a complete astronomically dated record of carbon and oxygen isotope ratio variations back to 66 million years ago. This means that the age of every geological event within that period that can be dated is now known with unprecedented accuracy. Westerhold and colleagues were able to use their astronomically tuned record to recognize four broad states through which climate oscillates, named the Hothouse, Warmhouse, Coolhouse, and Icehouse. Each depended on the orbital cycles, but also on the extent of reflective polar ice and the concentration of greenhouse gases in the atmosphere.

The main cause of the Milankovitch cycles is the gravitational pull of the Sun and the Moon on Earth, but astronomers recognize that the orbital motions of the other planets also induce terrestrial cycles. Venus and Jupiter, for example, cause the ~405,000-year cycle in Earth's eccentricity, which the huge mass of Jupiter stabilizes, allowing the cycle to remain constant for long periods of geologic time. Geologists began to look for this cycle in the rock record and found a number of candidates, but not until 2018 did they find one that had been independently dated by radiometric methods.

A group led by Dennis Kent of Rutgers University drilled into an ancient lakebed beneath the Newark Basin in New Jersey, where the rocks date to the Late Triassic and Early Jurassic periods, about 220 million years ago. The cores reveal a repeating cycle of sedimentary beds some 60 meters thick, with the boundary between them defined by unusually dark layers of shale. Paleomagnetic and radiometric dating showed that this pattern corresponded to the ~405,000-year cycle. The dark beds of shale in the Newark Basin cores formed in deep lakes under warm and humid conditions at times of high orbital eccentricity. They are separated by shallow lake deposits from periods when the climate was colder and drier, reflecting times of low eccentricity.

As one example of the benefits of an astronomical timetable, the 2012 version of the official geologic timescale had reflected an uncertainty in the age of a Late Triassic geologic boundary by listing two possible ages: 221.0 or 228.4 million years. The 405,000-year-old "metronome" of the astronomical timescale was used to resolve the discrepancy and establish the age of the boundary at 227 million years.[7]

Cyclostratigraphy

Long before scientific ocean drilling and astrochronology, geologists had routinely observed sedimentary rock strata in repeating cycles. In 1895, the great American geologist Grove Karl Gilbert (1843–1918) published an article titled "Sedimentary Measurement of Cretaceous Time."[8] He had studied "regular alternations of strata" at the base of the Rocky Mountains. One example was the Niobrara Formation, in which the cycles of harder chalk and softer and more easily eroded marl (a lime-rich mudstone) are impossible to miss.

At a time when the scientific community had turned its back on Croll's astronomical theory, Gilbert had the perspicacity and confidence to attribute the cyclical pattern to "the precession of the equinoxes." He thought that the known 21,000-year precession cycle was consistent with the width of the alternating strata and reasonable estimates of sedimentation rates. On that basis, Gilbert calculated that the "3900 feet (1200 m) of sedimentation [in the Niobrara] required about twenty million years"—a rate of about 6 cm per thousand years. In 2008, 113 years later, a modern study of the Niobrara using computerized astronomical tuning calculated a sedimentation rate of about 1.4 cm per thousand years, not far from Gilbert's figure.[9] At the time Gilbert wrote, according to Lord Kelvin, the entire age of Earth spanned only 100 million years, or five Niobrara formations.

Gilbert's theory that astronomical factors led to cyclic sedimentation lay dormant until Professor Alfred Fischer of the University of Southern California rejuvenated it in a 1986 article titled "Climatic Rhythms Recorded in Strata"—and launched a new field of geology called cyclostratigraphy.[10] His work took the evidence for astronomical causes of sedimentary cycles from deep-sea cores to outcrops on land, like those of the Niobrara. Cyclostratigraphy has allowed scientists to extend the astronomical timescale back beyond the reach of deep-sea cores and to accurately date much older rocks, as in the Newark Basin lakebeds.

It is no exaggeration to say that astrochronology and cyclostratigraphy represent scientific revolutions. They have led to a reconstruction and replacement of the standard geologic and paleomagnetic timescales; a better understanding of the intertwined evolution of Earth and the Moon; the recognition that other planets perturb Earth's astronomical cycles; the detection of Milankovitch cycles on Mars; an improved understanding of ancient solar system dynamics; and an improved understanding of how Earth's length of day has changed throughout geologic time. Let us take a closer look at two of these as examples.

When a smaller object orbits another, as the Moon does Earth, the smaller deforms the larger, as evinced by our ocean tides. The daily deformation of the larger object slowly dissipates energy, causing its rotation rate to slow. In order to conserve momentum, the smaller body recedes from the larger. Eventually the smaller takes the same amount of time to rotate on its own axis as it does to make one revolution around the larger, which is why the Moon always presents the same face to us.

The rate of Earth's rotation defines the length of the day. Since Earth rotated faster in the past, a day must have been shorter and a year must have had more of them. To us, this gravitational slowing would make it appear that the Moon was speeding up. Edmond Halley (1656–1742) of cometary fame was the first to notice this effect, based on his study of ancient eclipse observations.

Geologists have been able to estimate length of day from fossil shells of mollusks that grew so rapidly that they laid down one growth ring per day. A classic study from 1963 found that 65 million years ago there were 371 days per year, rising to 424 days about 600 million years ago.[11] This was a remarkable finding for its time, but understandably imprecise. Cyclostratigraphy offered a way to make much more precise measurements of the length of day over time. To take one example, in 2020 a Danish group located a pronounced obliquity cycle in sedimentary rocks near the Cambrian-Ordovician boundary that date to about 493 million years ago.[12] This allowed them to calculate the distance between Earth and the Moon for that time at 370,000 km (vs. 385,000 km today), the length of day at 21.8 hours, and the number of days per year at 402. The estimate from the fossil mollusks had been 412 days per year, not that far off.

Mars

Since the Milankovitch cycles are astronomical, they should be present on any planet where material has been deposited in layers over time, which essentially means as waterborne sediment. To turn this around, any planet that displays sedimentary layers with Milankovitch cyclicity had water on its surface in the past. The most accessible candidate is Mars and by now the many space missions to the Red Planet have found abundant evidence of the past presence of water. Figure 12.3 presents a clear example.

The Arabia Terra region in the north of Mars is one of the planet's most densely cratered and therefore one of the oldest. The image in figure 12.3 shows beds and bundles of beds within Becquerel crater, named for the discoverer of radioactivity. The beds are about 10 meters

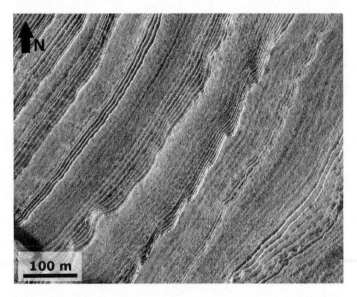

Figure 12.3
Mars cyclicity from the High Resolution Imaging Science Experiment (HiRISE), showing layered, cyclical strata within the wall of Becquerel crater. These layers were deposited in water.
Source: NASA HiRISE, in Kevin W. Lewis et al., "Quasi-Periodic Bedding in the Sedimentary Rock Record of Mars," *Science* 322, no. 5907 (2008): 1534.

thick and are bundled into larger groups, much like the sapropels on Sicily shown in figure 12.2. With no way to date the rock layers, it is not possible to match them to the astronomical cycles. That said, the scientists who have studied these images have made an educated guess that the repetition of the layers reflects the obliquity cycle on Mars and that they were deposited over a period of about 12 million years. The regularity of these cyclic sediments suggests a period of quietude that contrasts with the violence shown by the many impact craters, lava flows, and flood deposits seen on Mars. Perhaps during one of those quiet periods, life could have begun on the Red Planet, evolved, and even have left fossil evidence of its one-time presence. Thus, we see, and will see again, how scientific ocean drilling has taken scientists from the depths of the oceans out into the solar system.

13
Crater of Doom

At many times and places throughout history, people have come across huge bones unlike those of any known creature. Early in the nineteenth century, Frenchman Georges Cuvier explained that these were the remains of animals that no longer existed—that had gone extinct. In 1842, Richard Owen, an English paleontologist, coined the name "dinosaur" (from the Greek for "terrible lizards") for these mysterious, vanished creatures and founded the Natural History Museum in London to house their bones and promote their study. The question, "What killed the dinosaurs?" became one of the greatest scientific mysteries and would remain so for nearly 150 years.[1]

The Alvarez Theory

Throughout the decades, scores of theories were proffered to explain the death of the dinosaurs, but none won acceptance.[2] Then in 1980 came an article in *Science* titled "Extraterrestrial Cause for the Cretaceous-Tertiary Extinction."[3] The authors were the father-son team of Luis and Walter Alvarez, together with two colleagues, Frank Asaro and Helen V. Michel. Luis was a Nobel Prize–winning physicist, while Walter was a professor of geology at the University of California at Berkeley. They had set out to measure the length of time represented by a thin layer of clay that marks the geological boundary at which the dinosaurs went extinct. (Formerly called the Cretaceous-Tertiary or K-T boundary, now called the Cretaceous-Paleogene or K-Pg. See figure 18.1.) They thought that the amount of time the clay layer represented could reveal whether the dinosaurs had disappeared slowly or suddenly, providing a clue to the cause of their extinction. The method the Alvarezes chose was to measure the amount of the rare metal iridium in the boundary clay. Iridium is much more abundant in meteorites than in earthly rocks and as meteorites travel through the atmosphere, they shed a microscopic, iridium-bearing dust that settles down to Earth at a steady rate. Therefore, the more iridium in the clay layer, the longer it took to form. The Alvarez team expected to find a modest but detectable amount of iridium, but instead discovered that the element was enriched 30-fold in the boundary clay compared to the background amount. The cause of dinosaur extinction and the clay layer, they proposed, was the impact of a mountain-sized asteroid. Enriched iridium was found at many other K-Pg impact sites, as was quartz that had

been shocked at high pressure and glassy microspherules. These became diagnostic indicators of meteorite
impact.

Then ensued one of the most vituperative controversies in the history of science. Impact specialists and many
geologists sided with the Alvarezes, while paleontologists
and some ultraorthodox geologists ridiculed and con-
demned the hypothesis, one calling it "codswallop" and
another writing that it was "pathological science" and a
"scam."[4] It became harder to object to the Alvarez Theory
after geophysicists Antonio Camargo and Glen Penfield
published their discovery of a large crater of K-Pg age buried under a half-mile of sedimentary rock off the northwest coast of the Yucatán Peninsula. It became known as
the Chicxulub crater, after a nearby small town.[5]

The confirmation of the Chicxulub crater in 1990
caused many to accept the Alvarez Theory at least tenta-
tively, though there were holdouts, particularly among
the paleontologists. Scientists began to search for places
where they could sample and study rocks of the K-Pg
boundary and immediately recognized that several scientific ocean drilling legs had recovered cores that included
it. Geologists quickly turned their attention to cores from
drilling sites in the Gulf of Mexico, where nearness to
ground zero meant that the effects of the impact should
be greatest. In a 1992 article, Walter Alvarez and colleagues studied cores from DSDP sites 536 and 540 which
lie about 90 km apart at the deep-water entrance to the
Gulf of Mexico between Florida and the Yucatán.[6] Both

holes had cored an interval that was at least approximately of K-Pg age and in that interval, Alvarez and his colleagues found sedimentary deposits that "appear to be the signature of a giant tsunami" as well the tell-tale glassy microspherules, shocked quartz, and elevated iridium. Another group studied a more widely spread set of DSDP and Ocean Drilling Program (ODP) sites in the Gulf of Mexico and the Caribbean. Near the K-Pg boundary, they found giant sedimentary gravel deposits consistent with the collapse of the continental shelf, which a large impact could have triggered.[7] The next step was to see if the K-Pg boundary and its distinctive set of impact markers could be found in cores collected from much farther away than the Yucatán ground zero. One site that fit the bill was DSDP 596, located roughly in the middle of the Pacific Ocean. A team from UCLA found that the boundary rocks in the core contained large amounts of iridium as well as shocked quartz.[8] In 1999, Dutch paleontologist Jan Smit reviewed the known deposits of ejecta from the Chicxulub crater.[9] He included 13 DSDP/ODP drill sites, many which had some combination of iridium spikes, shocked quartz, and exotic microspherules. Smit showed conclusively that the effects of the Chicxulub impact were worldwide and that the farther from the crater, the thinner the K-Pg boundary layer. Other sites were drilled to investigate specific aspects of the Chicxulub event, such as the anomalous change in carbon isotope ratios at the boundary.

To take stock of the Alvarez Theory on the thirtieth anniversary of its announcement, in 2010 41 authors published an article in *Science* titled "The Chicxulub Asteroid Impact and Mass Extinction at the Cretaceous-Paleogene Boundary."[10] They noted that some 350 K-Pg boundary sites worldwide had been studied, making them "the most intensively investigated deposits in the geological record." Cores that included the K-Pg boundary had been recovered at 49 scientific ocean drilling sites, inspiring over 8,800 scientific articles.[11]

The authors of the 2010 review article concluded simply: "The Chicxulub impact triggered the mass extinction." But to show that the controversy was far from over, 28 opponents signed a letter of rebuttal, rejecting the "simplistic extinction scenario" and calling for a much more complex sequence of events. Three years later, Paul Renne of the University of California at Berkeley and nine colleagues used glassy, extraterrestrial microspherules from the K-Pg boundary to date the impact event at 66.038±0.025 million years.[12] To date the geological boundary, they collected from the Hell Creek site in Montana, source of the finest *T. rex* specimens. Samples from closest to the iridium peak there gave 66.043±0.011 million years. Within the precision of the method, the ages of the impact and the extinction are identical. Either this is cause and effect, or one of the greatest coincidences in earth history.

The *coup de grâce* to the opposition came in a March 2019 article reporting on a K-Pg fossil site in North

Dakota. The authors had found a meter-thick section of sediment containing a jumble of fossilized trees going every which way, along with fish and other creatures.[13] Extraterrestrial microspherules were found embedded in the gills of fossil fish and in tree amber, along with a large iridium spike. An article on the finding in the *New York Times* had the title, "Fossil Site Reveals Day That Meteor Hit Earth and, Maybe, Wiped Out Dinosaurs."[14]

Chicxulub

At 150 km in diameter and 20 km deep, the Chicxulub crater is the second largest on Earth. The largest known terrestrial impact structure is the 300-km wide, 2-billion-year-old Vredefort crater in South Africa, shown in figure 13.1. The third largest is the 100 km, 1.85-billion-year-old Sudbury crater in Ontario. Neither the Vredefort nor Sudbury structures was suspected of being an impact crater until pioneering geologist Robert Dietz found telltale "shatter cone" structures indicative of impact at both. In craters this old, and on most terrestrial craters, the surface evidence of impact has long since been eroded away. But as shown in the spectacular view of the Vredefort crater from space, the rock structures, once far below ground zero but now exposed by erosion, reveal the bullseye of an impact crater.

The Moon, by contrast, has no air or water and no erosion, allowing scientists to study features unchanged even after billions of years. They find that lunar craters

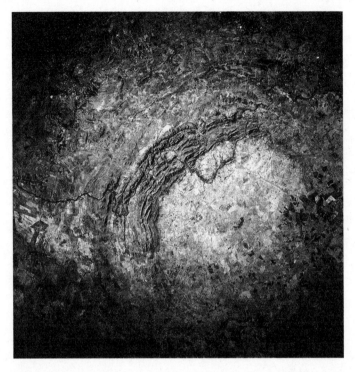

Figure 13.1
The 2-billion-year-old Vredefort structure in South Africa seen from space.
Source: NASA, Wikimedia Commons, August 29, 1985, https://commons
.wikimedia.org/wiki/File:Vredefort_Dome_STS51I-33-56AA.jpg.

larger than about 20 km in diameter have a central peak, like that of 80 km Tycho, the bright, rayed crater we see at the bottom left of the Moon and shown in figure 13.2. The size of craters with central peaks appears to relate to the gravity of the body on which they formed, ranging from roughly 30 km on Venus to 150 km on the Moon.

Figure 13.2
Tycho with its central peak.
Source: NASA, Wikimedia Commons, September 15, 2014, https://
commons.wikimedia.org/wiki/Category:Tycho_(lunar_crater)#
/media/File:Tycho_LRO.png.

In larger craters, an inner, broken circular structure of hills called a peak ring replaces the central peak. Craters with peak rings have also been imaged on Mars, Mercury, and Venus, so they must somehow be inherent to the formation of large impact craters. Lunar crater Schrödinger, shown in figure 13.3, at 500 km provides a good example.

Figure 13.3
Schrödinger with its internal peak ring.
Source: NASA, Wikimedia Commons, composite image, n.d., https://commons.wikimedia.org/wiki/Category:Schrodinger_(crater)#/media/File:Schr%C3%B6dinger_(LRO)_500_km.png.

Of course, no one can photograph the Chicxulub crater because it is underwater and buried below the floor of the Gulf of Mexico. But measurements of the pull of gravity over the area reveal the annular structure hidden below and make the broken peak ring obvious. Scientists have come up with two theories to explain the peak rings. One is that they form when an overly tall central peak collapses and its material flows outward toward the crater rim and piles up. The other is that peak rings represent deep, basement rocks uplifted by tens of kilometers of rebound immediately postimpact.

No crater exposed at the surface of Earth is large enough and sufficiently uneroded to display a peak ring, but the buried Chicxulub structure does. In the late spring of 2016, the International Ocean Discovery Program (IODP), the successor of the ODP, and the International Continental Scientific Drilling Program (ICDP) joined on Expedition 364 to drill the Chicxulub peak ring about 50 km from the center of the buried impact structure. Sean Gulick of the University of Texas at Austin and Joanna Morgan of Imperial College London were the cochief scientists/geophysicists of the expedition.

The mission planners estimated that the drill would reach the Chicxulub structure at about 1500 m below the seafloor. But the drill site was close to the Yucatán shoreline and too shallow to permit the operation of the IODP vessel *JOIDES Resolution*. Instead, the crew used a mission-specific platform named Lifeboat Myrtle, shown in figure 13.4. This was a structure used to service oil and gas wells in the Gulf with a deep-sea drilling rig welded onto the side, pictured here by the derrick. Drilling expeditions have used these platforms in a number of shallow-water settings, including near Tahiti and on the Great Barrier Reef. They are towed to the worksite, jacked up on the three legs to a desired height above the water, and typically drill a single hole. The platforms do not have

Figure 13.4
Chicxulub Expedition 364 mission specific platform: Lifeboat Myrtle.
Source: International Ocean Discovery Program.

scientific laboratories, so they are brought to the rig in shipping containers.

The drill core first brought up sedimentary rock, but then passed through broken-up (brecciated) rock containing fragments of impact melt, as well as shocked quartz and shatter cones, both dispositive evidence of impact. Sedimentary rocks typically accumulate at a rate of a few centimeters per thousand years, but Gulick noted that the core showed an accumulation of over 100 m in one day: the first day of the Cenozoic.[15] Then came the hoped-for payoff: "It was limestone, limestone, limestone, breccia. And then suddenly pink granite!" exclaimed Gulick. "It was exhilarating, it looks like your classic pink granite countertop."[16]

Granite is a rock of the deep crust. Its presence reveals that the Chicxulub impact was so powerful that it lifted to the surface abyssal basement rocks that before the impact had been at a depth of at least 7.5 km. Thus, the Chicxulub peak ring did not form by collapse of a surficial central peak, but by uplift from deep below. The process was almost instantaneous, as Gulick observed in an interview: "Unlike tectonic mountains [those on Earth] that take millions of years to form, these mountains [the peak ring] are made in less than 10 minutes."[17]

The discovery that peak rings come from great depth means that scientists can use them to provide information on the internal layering and composition of other planets and moons. For example, if future astronauts were to sample Schrödinger's peak ring, also presumably uplifted from great depth, scientists might learn a great deal more about the interior of the Moon without having to drill.

The Chicxulub cores have led to many important discoveries. One is that the asteroid's angle of approach made the effects of the impact worse. Imagine that you are trying to skip a flat stone across a pond, as many have done as children. If you were to drop the stone directly from above—90°—it would not skip at all but splash and sink. There is one angle that is just right to get the maximum number of skips from the stone. Just as with a skipping stone, an incoming asteroid has an optimum angle of approach that on impact will launch the maximum amount of vaporized rock. Gareth Collins and his colleagues ran computer simulations of the impact and found that the angle of approach of the Chicxulub impactor had been between 45° and 60°, causing it to release more vaporized material than either a shallower or steeper trajectory.[18]

Aftermath

Another factor that multiplied the effects of the Chicxulub impact was that the target rocks at ground zero contained calcium carbonate and sulfur, so that the impact ejected huge amounts of CO_2 and sulfur dioxide, much of which reached the stratosphere. Estimates are that the impact released ~425 gigatons (billon metric tons) of CO_2 and ~325 gigatons of sulfur. This is about one-fourth the total amount of CO_2 emitted by all human activities to date.[19]

We know from observation that volcanic eruptions put sulfur aerosols in the air. They reflect sunlight and cause global temperature to fall. (Some scientists are exploring the idea of injecting sulfur into the atmosphere to offset human-caused global warming, but the cure might be worse than the disease.) The Chicxulub impact would have put vastly more sulfur into the atmosphere than any number of volcanic eruptions, blocking sunlight and causing an "impact winter," analogous to the nuclear winter that scientists have feared could result from a global nuclear war.[20] Then, over the longer term, the added CO_2 would have raised temperatures again.

Scientists estimate that about 70 percent of species went extinct at the K-Pg. But the event was especially hard on the planktonic forams, which lost an estimated 87 percent of species. A geochemist named Michael Henehan came up with a possible explanation. In 2016, he attended a conference on paleoceanography at Geulhem in the Netherlands, known for its nearby cave dwellings. The participants made a field trip to the caves, which contain rocks close to the K-Pg boundary. As the story goes, Henehan came across a thick clay formation that just postdated the boundary.[21] Having brought no sample bags, he put his lunch in his pockets and used the lunch bag to hold specimens he collected from the layer. Back in his lab at Yale, Henehan found the layer filled with foram shells. And he knew just what to do with them: measure the ratio of the boron isotopes.

Scientists have discovered that the ratio of B-10 to B-11 is directly proportional to the acidity of seawater. Measuring the boron isotope ratio in forams from Geulhem and three DSDP and six ODP sites, Henehan and colleagues found that the acidity of the ocean surface waters increased drastically in the first 100 to 1,000 years following the Chicxulub impact.[22] This was due to the sulfur added to the atmosphere by the impact, which turned into sulfuric acid and rained down to decimate the forams.

Another important finding from the drilling of the Chicxulub crater peak ring is that superheated seawater may have remained in the shattered crust below the crater for up to a million years after the impact. This "hydrothermal system" might have resembled the one that we see today in Yellowstone National Park, where heat-loving (thermophilic) organisms thrive in brightly colored pools of hot water. On early Earth, the multitude of impacts could have created hydrothermal areas that incubated microbes and led to the early evolution of life. Geologist and Chicxulub specialist David Kring of Houston's Lunar and Planetary Institute has developed this idea into a hypothesis of the "impact-origin of life."[23] He and colleagues have noted that in the Hadean eon at the beginning of earth history, as many as 6,000 impactors larger than the ~10 km Chicxulub asteroid may have struck our planet. An estimated five impactors would have been larger than ~500 km and three larger than ~700 km. They would have created approximately 200 impact craters between 1,000 and 5,000 km wide, providing plenty of opportunity for thermophilic organisms to give life on Earth a start. It is a provocative thought that at the beginning of earth history, impact may have created the conditions for life, then four billion years later destroyed the dinosaurs and 70 percent of all living creatures, while allowing a small, squirrel-sized mammal to survive.

14
The Hot, Deep Biosphere

When HMS *Challenger* set sail in 1872, some scientists still believed in the azoic theory: that life cannot exist below 300 fathoms. Others thought that creatures lived in the abyss, but that the cold and dark prevented them from evolving. With no more than their dredges, the *Challenger* scientifics soon disproved both ideas. The exploration of life at and below the surface of the dark seafloor began with a 1936 article by Claude ZoBell and Quentin Anderson of the Scripps Institute of Oceanography, who found abundant bacteria in the surface layers of sediment cores 40 to 75 cm long taken off the coast of Southern California.[1]

The deep sea and its creatures became a subject of great interest in the 1930s, prompted by the invention of the deep-sea submersible, a sort of mini submarine built to withstand the great pressures of the abyss. The most notable of these early vessels was the two-person "Bathysphere" used by famed scientist and author William Beebe (1877–1962), whose books with their photos of bizarre deep-sea creatures fascinated and inspired youngsters of an earlier time.[2] Engineer Otis Barton designed the vessel and he and Beebe used it to make a number of deep dives off the coast of Bermuda. In 1934, the two reached a record depth of 923 m.

The successor to Beebe's Bathysphere was the Alvin, named for its inventor, the eponymous Al Vine, and launched in 1964 by Woods Hole Oceanographic Institution. It was designed to carry two scientists and a pilot 4,500 m down and allow them to stay at that depth for

Figure 14.1
William Beebe (*left*) and Otis Barton (*right*) with their Bathysphere.
Source: NOAA Ocean Exploration, courtesy of the Wildlife Conservation Society, https://oceanexplorer.noaa.gov/explorations/05stepstones/logs/aug15/media/WCS_Beebe_Barton.html.

nine hours. Alvin made over 5,000 dives and fostered an estimated 2,000 research publications. But it had a rocky start, to say the least. Alvin's first dive was in 1965 to 1800 m. In March 1966, Alvin was used in an unsuccessful attempt to recover a hydrogen bomb that had been lost in a midair accident and fallen to the seafloor at 910 m depth off the coast of Spain. Then in October 1966, as Alvin was being lowered over the side of its support vessel, with crew members aboard and the hatch open, the two steel cables holding it broke. The crew was able to escape, but the vessel fell to the seafloor in 1500 m of water. The fortunate crew members had left their lunches behind, and when Alvin was hauled up, there the food was, intact and with no sign of attack by snacking microbes. This reinforced the view that the deep sea was inimical to significant bacterial life.[3] The Alvin's most famous dive, however, was its 1968 exploration of the wreck of the ill-fated *Titanic*. After a complete overhaul, finished in 2014, Alvin was back on active duty at Woods Hole. In the summer of 2022, the submersible reached a record depth of 6,453 m in the Puerto Rico Trench, meaning that Alvin could reach nearly any point on the seafloor.

Black Smokers

One of the key concepts of plate tectonics is that underneath the center of the oceanic ridges lie chambers of molten magma. These heat the adjacent seawater, which rises and flows out via hydrothermal vents. In 1977, scientists made 24 dives in the Alvin to study these vents along the Galápagos Rift, an offshoot of the East Pacific Rise. They found that two-thirds of the heat lost at the rift escapes via these outlets.[4]

Prior to these dives in the Alvin, scientists believed that photosynthesis was the ultimate and indispensable source of the energy needed to support life, meaning that living creatures could not exist in the blackness of the ocean depths. Yet the scientists aboard Alvin found abundant life in a variety of forms on the Galápagos Rift at a depth far below that to which sunlight can penetrate. Where did these creatures get the energy to sustain themselves if not from photosynthesis? Scientists had answered that phytoplankton, which consists of microscopic plants and lives near the surface, die and sift down as "marine snow," a term coined by Beebe and a process recognized by the *Challenger* scientifics. Rachel Carson described it in *The Sea Around Us*: "When I think of the floor of the deep sea . . . I see always the steady, unremitting, downward drift of materials from above, flake upon flake, layer upon layer—the most stupendous 'snowfall' the Earth has ever seen. . . ."[5] The dead phytoplankton fall to the dark, abyssal ocean floor and provide a food source for organisms living there. According to this theory, photosynthesis would still be the ultimate energy source for the creatures of the abyss.

But the hydrothermal vents on the Galápagos Rift held a concentration of organisms thousands of times greater than the seafloor around them. Some unrecognized process within the vents—not photosynthesis—was providing the energy on which the vent ecosystem depends. It turned out to be "chemosynthesis," in which bacteria oxidize inorganic materials, primarily hydrogen sulfide, in chemical reactions that in turn provide the energy to sustain higher life forms. Black, irregular chimneys mark some vents where chemicals that had been dissolved in the hot water have precipitated as dark sulfides when the hot vent water meets the cold ocean. Scientists subsequently found these "black smokers" in many places in the Atlantic and Pacific Oceans, as well as stranded on land—for example, along the California coast where plate tectonics has lifted an old ocean floor well above sea level.

The creatures of the vent ecosystems ultimately depend on the sulfur-reducing bacteria—which we could almost say "breathe" sulfur—and include many strange denizens never before seen. None were stranger than the tube worms, which measure up to 3 m long but are only 4 cm wide and live in clusters of thousands of individuals per square meter. They depend on the bacteria for energy and have no need for a digestive system. Their existence in such inhospitable conditions once again raised questions about what other life forms could exist at and below the seafloor.

Thomas Gold

If life can exist in the depths of the ocean, could a significant portion of all life on Earth be in those depths, rather than above them? That was the thesis of one of the most inventive and iconoclastic scientists of the second half of the twentieth century. Thomas Gold was born in Austria in 1920 to Jewish parents who fled to England in 1938 after Hitler annexed Austria. Gold entered Trinity College at Cambridge, but when World War II broke out the British interned him as an enemy alien and deported him to a camp in Canada. After 15 months there, he was allowed to return to England, where he reentered Cambridge to study physics and worked on the all-important radar. Gold's multifarious interests and accomplishments are enough to fill a book, or several. It is no surprise that he would be one of the first to explore the larger implications of the deep hydrothermal vents.

In a provocative 1992 article, "The Deep, Hot Biosphere," and in a 1999 book of the same title, Gold extrapolated from the microbial life of the vents to propose that such life also existed in abundance beneath the seafloor.[6] He went so far as to suggest that subsurface microbial life could be comparable in mass and volume to all life on the surface. Microbial life could be pervasive everywhere beneath Earth's surface in the pore spaces between mineral grains—and not only on

Earth but also on other bodies in the solar system: the Moon and Mars, for example. They have too little air and water to sustain life on their surfaces, but it could well exist below. Perhaps microbial subsurface life came first, protected from the surficial violence of the early solar system and using chemosynthesis, then evolved into photosynthetic life. Gold thought that microbial life might be widespread in the universe, a concept known as panspermia which goes back to the Greek philosopher Anaxagoras in the fifth century BCE. Many notable scientists have endorsed the idea, but with no way to test it, attention shifted to the possibility that the organic building blocks of life might have been present throughout the solar system at its beginning.

Drilling the Abyss

The detection of life beneath the seafloor was the goal of one of the earliest DSDP voyages, Leg 15 in 1970, led by chief scientist Wallace Broecker of Columbia University. The crew found methane, a byproduct of microbial activity, in sediments 800 m beneath the seafloor and tens of millions of years old.[7] In October 1986, the crew of DSDP Leg 96 drilled the Mississippi Fan, a submarine pile of sediment in the northeastern Gulf of Mexico. They found subsurface microbial activity down to 167 m beneath the seafloor. By the end of the century, the Ocean Drilling

Program had sampled 14 sites for evidence of bacterial activity. A summary of these studies found that although the number of microbes typically decreases with depth beneath the seafloor, living cells are still present down to 700 m. The authors came to the remarkable conclusion that the biomass in the top 500 m of seafloor sediments equals 10 percent that of total surface biosphere.[8] These early results suggested that living bacteria likely exist at greater depths than drilling had yet reached. This led to the first expedition designed specifically to study subsurface life.

In the spring of 2002, ODP Leg 201 drilled in two locations, one on the continental margin off Peru and the other in the equatorial Pacific. The subsurface ecosystems turned out to have a great diversity of microbes, including not only the sulfate-reducing bacteria found at the vents but a new type that got its energy from carbon reactions. The microbes were "alive" in that they engaged in metabolic activities such as repairing DNA and undergoing cell division. They included all three domains of life: archaea (one-celled organisms), bacteria, and eukaryotes (cells that have a nucleus). By this time, scientists estimated that subsurface bacterial life could amount to one-third that of Earth's total living biomass.[9] In 2003, ODP Leg 210 drilled the seafloor off Newfoundland and upped the ante once again. It found living bacterial cells 1,626 m below the seafloor, in rocks 111 million years old, at temperatures of 113°C. This led the authors to estimate that bacteria in subsurface sediments may make up as much as two-thirds of total bacterial biomass.[10]

Expedition 329 of the Integrated Ocean Discovery Program (IODP), which followed the ODP, in October 2010 drilled in the South Pacific Gyre, some of the deepest water on Earth. It is the largest of the five giant oceanic systems of rotation that move enormous volumes of seawater. The South Pacific Gyre rotates counterclockwise, bounded by the equator to the north, Australia to the west, South America to the east, and the Antarctic Circumpolar Current to the south. Its center is the "oceanic pole of inaccessibility": the location farthest from any continent. The South Pacific Gyre has one of the lowest sedimentation rates in the oceans and its bottom sediments have the lowest cell concentrations and the least metabolic activity of any. To discover the most extreme conditions under which life can exist on Earth, this is the place to go.

Aboard the *JOIDES Resolution*, still hard at work after all these years, in water nearly 6 km deep, the scientists drilled 100 m into the seafloor. They found microbes all the way to the bottom of the cores, but at an abundance of only about 1,000 cells per cubic centimeter of sediment, compared with 100,000 or more in a typical nutrient-rich sediment from the ocean margins. The scientists estimated that the deepest microbes were at least 100 million years old, making it seem they could only be fossils. Surely nothing could "survive," whatever that means exactly, for 100 million years. But when brought back to the lab and offered nutrients, the microbes began to grow. After being fed for 65 days, their abundance had increased to 1 million cells per cubic centimeter. One of the leading authorities on microbial life in the seafloor, Steven D'Hondt of the University of Rhode Island, said, "What's most exciting about this study is that it shows that there are no limits to life in the world's ocean. In the oldest sediment we've drilled, with the least amount of food, there are still living organisms, and they can wake up, grow and multiply."[11]

This seemingly fantastic discovery raised the question of what the microbes beneath the gyre had been doing for 100 million years. Perhaps the cells had too little food to divide, but enough to repair damaged molecules. But that "seems insane," said D'Hondt, who wondered whether there is not another undiscovered source of energy—possibly radioactivity—that could support slow cell division.[12]

On Expedition 337 of the IODP, the Japanese drilling ship *Chikyū* (Earth), designed for deep-sea drilling, cored to a depth of 2,466 m beneath the seafloor off Japan's Shimokita Peninsula. It found microorganisms in coal and shale that resemble those in the soil of modern tropical forests. These microbial communities are thought to be relics of those that inhabited soils about 20 million years ago, rather than more modern microbes that might have migrated into the coal layers from elsewhere. To explore the upper temperature limit at which microbes can survive, on Leg 370 of the IODP, *Chikyū* drilled in the Nankai Trough subduction zone off Cape Muroto in south-central Japan.[13] The drill reached 4,776 m and the deepest core was collected at 1,177 m, where the temperature measured 120°C. Microbial life was detected all the way to the bottom of the sediment column, though the number of cells per cubic centimeter was extremely low, as would be expected. The cells at that depth appeared to spend most of their energy repairing the damage caused by the high temperature. Several authorities had written that the temperature limit to life in the subsurface was 80°C, but Gold had predicted that the upper temperature limit on bacterial life would be in the range of 120°C to 150°C—and he turned out to be right. Indeed, the results of Leg 370 show that one-fourth of the sediment on the seafloor is buried where temperatures are above 80°C.

Martians

These findings from scientific ocean exploration suggest that microbial life may be pervasive everywhere beneath Earth's surface under conditions long thought to be inhospitable, if not fatal. This raises the possibility that, as Gold postulated, bacterial life may have existed and may still exist on other bodies in the solar system, including Mars. This despite the Red Planet's hellish surface conditions, continually blasted by lethal radiation from the Sun and the cosmos. The surface temperature of Mars averages −60°C and is so dry that a cup of water would vaporize instantly. A group of scientists experimented with a terrestrial bacterium called *Deinococcus radiodurans*, said to be the toughest on Earth according to Guinness World Records, to test whether it could survive on Mars.[14] This creature thrives in nuclear reactors. They found that if buried 9 m underground, *D. radiodurans* could withstand Martian levels of radiation for 280 million years. Later this decade, the European Space Agency plans to send a spacecraft to Mars that will drill more than 2 m below the surface and analyze the organic molecules found there. How will we humans react if, when life is discovered on another planet, it looks nothing like us, nor even little green men, but is microbial? If evidence of microbial subsurface life is found on Mars, it may have been the first lifeform in the solar system, where, protected from the surficial violence and using chemosynthesis, it could indeed have evolved into photosynthetic life and eventually, us. If Earth and Mars harbor subsurface bacterial life, why not other planets as well?

15
The Seventh Continent

Antarctica, the last continent to be discovered, is the most inaccessible, coldest, driest, and windiest. It has an area of 13.7 million sq km, well surpassing the nearly 10 million sq km of the continental US. The ancients had long suspected that Earth had a seventh continent, out of sight somewhere far to the south. An early Greek cartographer, Marinus of Tyre, named it Antarctica since it had to be oppositely situated from the Arctic. By the seventeenth century, mariners had rounded both Cape Horn and the Cape of Good Hope, suggesting that Antarctica was likely surrounded by water and was large enough to qualify as the seventh continent. In 1773, Captain James Cook in HMS *Resolution* crossed the Antarctic Circle at 66°S, not quite close enough to spot the continent. That distinction is credited to Russian Admiral Fabian Gottlieb von Bellingshausen, who in circumnavigating the globe in 1820 ventured within 30 km of the long-suspected land mass, close enough to discern it.

One could make the case that in its multiple effects on the rest of Earth, Antarctica is the most important continent. Its vast ice sheet reflects so much sunlight that it helps determine global temperature and climate. Antarctica is so large and so located that it plays a crucial role in oceanic circulation. Encompassed by the unbroken Antarctic Circumpolar Current (ACC), the largest ocean current globally, its circulation dominates the Southern Ocean, linking to the Atlantic, Pacific, and Indian Oceans. The oceanic conveyor belt carries deep North Atlantic water far to the south, where the ACC distributes it to the other ocean basins, testament to Antarctica's pivotal role in overall oceanic circulation.

Today, as shown in figure 7.5, the Antarctic Plate borders the African, Australian, Pacific, Nazca, and South American Plates, so that it has played a critical role in the plate tectonic evolution of Earth. Off the coast of Chile, the Antarctic Plate, South American Plate, and Nazca Plate meet at a single point, called a triple junction. Many triple junctions occur where three mid-oceanic ridges meet, but the Chile Triple Junction is unusual in that here a ridge is directly subducted beneath a plate, with the Peru-Chile Trench marking the point of descent. Geologists have learned a great deal about how plates move and interact by studying these triple junctions.

Prior to the Triassic period 200 million years ago, Antarctica had been part of the supercontinent Pangaea. The Atlantic Ocean began to open about 175 million years ago, splitting Pangaea into Laurasia in the north and Gondwana in the south. Instead of the frozen polar desert that we know today, at that time Antarctica was forested, with warm-water ammonites near shore and reptiles and many other species on land. From then Antarctica moved ever more southward toward the pole, by the Late Cretaceous arriving at its present location and with roughly the same size and shape as today.

Drilling in Antarctic Waters

Antarctic ice comprises nearly three-fourths of the world's freshwater stock and were all of it to melt, sea level would rise nearly 60 m. This may seem unimaginable, but for much of its history, Antarctica was ice-free. The water that would become Antarctic ice was then stored in the oceans, causing sea level to be much higher. The geological history of Antarctica, and in particular when and how it became 98 percent ice-covered, is a matter of strict importance—and not only to geologists. It is no exaggeration to say that humanity's future depends on preventing massive melting of polar ice. With a present-day atmospheric CO_2 concentration of 420 ppm, the tipping point that would induce dangerous melting of Antarctic ice is close, if it has not already been crossed. Certainly at 500 ppm, which business-as-usual will bring by 2050, it may be too late to prevent dangerous melting of the polar ice caps.

Given the importance of the geological history of Antarctica to plate tectonics, world climate, oceanic circulation, and sea level, it became an early target of scientific ocean drilling. During the Austral (southern hemisphere) summers between 1972 and 1976, the DSDP sent four expeditions to the Antarctic. Leg 28 departed Fremantle, Australia in late December 1972, accompanied by two ice-breakers, one of which had to do double duty on arrival by pushing away approaching icebergs. It would have been hard to find a more forbidding place to drill, yet the dynamic positioning system kept the ship on hole even in 40-knot winds and two-meter swells. Spectacular results were obtained on the twin problems of the plate tectonic evolution and paleoclimate of Antarctica.[1]

One of the main questions the early deep-sea drilling legs hoped to answer was when glaciation began in Antarctica. With no reason to think otherwise, geologists had assumed that the continent had begun to glaciate at the onset of the Pleistocene, the epoch of the ice ages, about 2.6 million years ago. But how do geologists study past glaciation? The ice has melted and vanished, so they rely on the presence of deposits the glaciers have left behind. These giant ice sheets grind up whatever they travel over, pick up the resulting debris, freeze it into the ice, then release it when the ice eventually melts, which may happen hundreds of kilometers away. Glaciers flowing off a land surface into the sea "calve" icebergs that are then caught in currents and carried away, melting along the way and dropping the debris that had been frozen within them onto the seafloor. Geologists recognize such "ice-rafted" debris by the diagnostic scratches on the surfaces of these "dropstones," among other evidence. They can date the debris in drilling cores from the accompanying microfossils. The surprising result from Leg 28 was that extensive glaciation of Antarctica began at least by the Early Miocene, 20–25 million years ago, much earlier than had been assumed.[2] The antiquity of Antarctic ice, and its variability, surely have to be counted as some of the most important findings of scientific ocean drilling.

To start Leg 29, *Glomar Challenger* sailed on March 2, 1973 from Lyttelton, New Zealand into the Southern Ocean, with microfossil specialist James Kennett as cochief. Sixteen holes were drilled in an area bounded by Australia and Tasmania, New Zealand, and Antarctica. Kennett and shipmates separated the forams from the sediment cores and sorted them into surface and deep-water species. Kennett sent a suite to Nicholas Shackleton at Cambridge, introduced in chapter 11, for analysis of the oxygen isotope ratios of the forams, a proxy for the temperature of the water in which they lived. Again, the result was a surprise: the oldest forams dated to the Late Paleocene, about 55 million years ago, and showed that the surface water temperature at that time had been nearly 20°C. Surrounded by water that warm, Antarctica was surely not covered by an ice cap. From there, the oxygen isotope ratios of the forams declined steadily to correspond to today's 0°C, indicating that Antarctica had gotten steadily colder for all those millions of years.

In a landmark 1977 article, Kennett summed up what had been learned from the four early drilling legs about the evolution of Antarctic glaciation, its encirclement by unbroken ocean, and their impact on global paleocean-ography.[3] First, he noted how the magnetic lineations found by Leg 28 had revealed that New Zealand had broken off from Australia about 80 million years ago. At the beginning of the Cenozoic Era, 66 million years ago, Antarctica had already moved well to the south, but was not yet glaciated. Kennett concluded that the absence of ice at this time showed that proximity to the poles had been insufficient to induce continental glacia-tion in Antarctica—instead, the obstructing land masses had to move away, allowing unhindered ocean circu-lation around the continent. About 55 million years ago, Australia broke away from Antarctica and drifted northward at about 5 cm/year, opening the Tasman Sea between Australia and New Zealand. At this time Ant-arctica may have had small mountain glaciers but had not yet developed a continental ice sheet. The end of the Eocene and transition to the Oligocene about 34 mil-lion years ago saw dramatic change. The tropical faunas and floras gave way to species that could tolerate colder temperatures, and sea ice began to form around Antarc-tica. The ocean conveyor belt became established at this time. The calcium carbonate compensation depth—at which acidic seawater dissolves calcareous shells, first noted by the *Challenger* scientifics—deepened rapidly. By 35 million years ago, Australia and Antarctica were nearly 1,000 km apart, separated by a deep-water chan-nel. The Drake Passage between Antarctica and South America may have opened at this time, initiating the

Antarctic Circumpolar Current. Kennett proposed that this isolated Antarctica so that as it moved farther south, becoming steadily colder, an ice cap was able to form. Later climate modeling suggested that a decline in atmospheric CO_2 concentration may have also influenced the temperature decline.[4]

In sediments dated to the Oligocene, about 25 million years ago, scientists found abundant ice-rafted debris in cores from the Ross Sea, a large embayment to the south of Antarctica, evidence that glaciation had begun on the continent by then. By the mid-Miocene, between 14 and 11 million years ago, a semipermanent ice cap had developed on Antarctica, where it has remained since, expanding and contracting. About five million years ago, during the latest Miocene, a global climatic cooling caused the volume of the ice cap to expand even beyond what we find today. Between 3 and 2.5 million years ago, ice sheets began to develop in the northern hemisphere, paced by the orbital cycles, to launch the Pleistocene epoch.

Antarctic Warming Episodes

In addition to the four DSDP legs, both the ODP and the IODP sent several expeditions to drill in Antarctic waters. For us today, the most important result of these expeditions is what they reveal about the episodes of global warming that interrupted the overall prolonged Antarctic cooling. One such period was the Miocene Climatic

Optimum between about 18 and 14.8 million years ago. Temperatures averaged 3–4°C higher than today, making this the warmest interval in the last 35 million years.[5] The Antarctic Drilling Project (ANDRILL), based at Antarctica's McMurdo Station, drilled from a floating sea-ice platform in the western Ross Sea.[6] Analysis of the core showed that the Miocene climate of the Antarctic coast was quite variable, changing from warmer to cooler in sync with the orbital cycles, but also in response to atmospheric CO_2. During the relatively brief warm periods, surface air temperature rose to 10°C; tundra extended as much as 80 km inland; surface water temperature rose; and the Antarctic ice sheet retreated inland, raising sea level by as much as 35 m, which could only have resulted from extensive melting of the ice sheet. When the climate cooled, these trends reversed and the ice sheet advanced into the sea and across the continental shelf. The maximum ice sheet advance occurred when the level of atmospheric CO_2 was around 280 ppm, the preindustrial concentration, and the maximum retreat when it was around 500 ppm.

Another warm period occurred during the mid-Pliocene, 3.3 to 3 million years ago, when global temperatures were 2°C–3.5°C warmer than today. The West Antarctic Ice Sheet (WAIS) periodically deglaciated, as did parts of the East Antarctica Ice Sheet (EAIS). Nearly all of the Greenland Ice Sheet also melted. Scientists estimate that sea level may have risen by 30 m: 22 m from the melting of the Antarctic ice sheets and 7 m from the Greenland Ice Sheet.[7]

During what geologists call Marine Isotopic Stage 11 of the Pleistocene, dated to ~400,000 years ago, global temperature was 1.5°C to 2.0°C warmer than preindustrial times and sea level was 6–15 m higher than today. This rise required the loss of most or all of the ice on the WAIS and the Greenland Ice Sheet, plus enough melting of the EAIS to combine to add 5 m to the sea level rise.

Antarctica and Future Sea Level Rise

We have seen that past melting of Antarctic and Greenland ice has caused sea level to rise and fall by meters and even tens of meters. Our concern is how much sea level may rise in this century and beyond. For that, we have the authority of the sixth annual report of the Intergovernmental Panel on Climate Change (IPCC), published in 2021.[8] As shown in figure 15.1, the IPCC used five different emissions scenarios to project sea level rise to the end of the century. IPCC scenario SSPS 8.5 has the highest emissions of any, but is also the business-as-usual emissions trajectory that the planet has been on. To get a sense of the worst impact that climate change might have, we must assume that emissions remain on the SSPS 8.5 path. Under that scenario, by the year 2100, temperature will have risen by 5°C and sea level from 0.6 to 1 m. However,

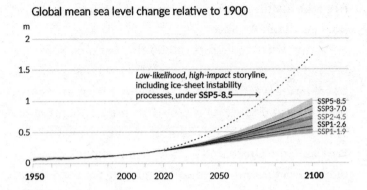

Figure 15.1
IPCC scenarios for sea level rise.
Source: IPCC, "Summary for Policymakers," in *Climate Change 2021: The Physical Science Basis. Contribution of Working Group I to the Sixth Assessment Report of the Intergovernmental Panel on Climate Change* , ed. V. Masson-Delmotte et al. (Cambridge: Cambridge University Press, 2021), 22, doi:10.1017/9781009157896.001.

the IPCC notes that a rise of 2 m "cannot be ruled out." By 2150, the likely sea level rise is from 1 to 1.9 m, but a rise of as much as 5 m is possible. Looking further ahead, sea level could rise 7 m by 2300, and the IPCC could not rule out a rise of 15 m. It is hard to see how human civilization could survive. As if that were not a chilling enough thought, the IPCC writes, "Many changes due to past and future greenhouse gas emissions are irreversible for centuries to millennia, especially changes in the ocean, ice sheets and global sea level." In other words, on our timescale, global warming is forever.

16
Eocene Doomsday

From its lack of glacial deposits and other evidence, geologists have long known that the Eocene epoch (56 to 34 million years ago) was unusually warm. Thus, the transition from the preceding, cooler Paleocene epoch to the Eocene marked the onset of global warming, the very problem that confronts humanity today. The Paleocene-Eocene Thermal Maximum (PETM) saw global average temperatures rise by an average of 6°C. If the Paleocene-Eocene (P-E) boundary could be sampled adequately, this real-world example of global warming would help us better understand the danger we face.

The Paleocene-Eocene Thermal Maximum

On January 5, 1987, the *JOIDES Resolution* sailed on Leg 113 of the ODP from Punta Arenas, Chile, bound for the Weddell Sea, the large embayment to the east of the Antarctic Peninsula. The crew drilled nine holes, one of which engendered hundreds of scientific articles and demonstrated the effects of global warming more dramatically than any computer model. The information gleaned from Hole 690B stands as a shining example of the value of scientific ocean drilling. As of June 2023, the definitive article on the analysis of the core by cochief scientist James Kennett, whom we met in the preceding chapter, and coauthor L. D. Stott has been cited more than 1,600 times.[1]

Kennett and Stott found that average global temperature had increased in the Early Eocene by about 6°C, reaching its highest point since the extinction of the dinosaurs 66 million years ago. The change took place in less than 20,000 years and the whole event lasted about 200,000 years.[2] Warmer water can hold more CO_2, which converts into carbonic acid, one of the many deleterious effects of global warming. At the P-E, as at the K-Pg, the oceans acidified and also deoxygenated. The calcite compensation depth rose sharply, showing that the water had absorbed a large amount of CO_2, and sea level rose 12–15 m higher. The bottom-dwelling (benthic) forams suffered their greatest extinction in the last 90 million years. On land, many organisms evolved to smaller sizes (dwarfing) and mammals—including primates—made their first appearance.

To understand the timing of the PETM required that it be accurately dated, but at about 55 million years of age, it lay just beyond the approximately 50-million-year limit of the astronomical timescale at the time. ODP Leg 198 to the Shatsky Rise in the west-central Pacific and

Leg 208 on the Walvis Ridge off the southwest coast of
Africa in the Atlantic later collected excellent sections
across the P-E boundary, which Thomas Westerhold and
colleagues used to identify the ~400,000-year eccentricity cycle in these cores. This "floating" chronology was
not connected to the rest of the astronomical timescale,
but still allowed the duration of the Paleocene epoch to
be measured at 24 eccentricity cycles, or 9,720,000 years.[3]
By 2019, calculation methods had progressed enough to
allow scientists to add the period from 53 to 58 million
years to the astronomical timescale. This dated the onset
of the PETM to 56.01±0.05 million years and its duration at 170,000±30,000 years.[4]

Possible Causes of the PETM

Despite the nearly four decades that have passed since
the drilling of Hole 690B, and despite the intense study
of the PETM ever since, scientists have been unable
to settle on its cause. To narrow the possibilities, they
first estimated how much CO_2 had been added to the
atmosphere at the P-E. Potential sources that were not
capable of adding that much carbon could be ruled
out. Their calculations show that at the P-E boundary,
between about 2,000 and about 7,000 gigatons (billion metric tons) of carbon were added to the oceans,
an amount that spans the carbon emissions that would
result were humans to burn all known reserves of coal,

oil, and natural gas. By one estimate, CO_2 levels during the PETM reached 1790 ppm.[5]

Scientists have offered a number of potential sources for the carbon injected into the atmosphere to launch the PETM:[6]

Methane clathrates: Microbes in deep-sea sediments continually produce methane gas, whose molecules can be trapped in a cage of ice. If these biogenic "clathrates" had melted at the PETM, they would have released into the atmosphere large amounts of methane, a much stronger greenhouse gas than CO_2.

Wildfires: A drier and more oxygen-rich climate could have led to the burning of peat and coal, releasing carbon into the atmosphere.

Volcanic activity: The North Atlantic experienced increased volcanism just prior to the PETM. Intrusion of hot magmas could have metamorphosed sedimentary rocks, releasing vast amounts of carbon and sulfur dioxides.

Meteorite impact: At drill sites on the continental shelf near New Jersey and another site 1,100 km away on the submarine Blake Plateau off the coast of Florida, scientists have found extraterrestrial microspherules and other strong evidence of meteorite impact at the P-E boundary. The microspherules date to the same age as the boundary, showing that they were not recycled from older impacts, such as the one that created the Chicxulub crater. The scientists who discovered this evidence do not claim that an impact event directly caused the effects at the P-E boundary. Rather they posit that the impact triggered additional volcanism in the North Atlantic.

Kimberlite eruption: Kimberlite is a rare igneous rock and the source of the world's diamonds. It is explosively injected into the upper crust from deep within the mantle, often propelled by CO_2. A large cluster of kimberlite pipes found in northern Canada date to ~ 56 million years, the same age as the PETM. Perhaps this event raised CO_2 levels enough to start the PETM.

Permafrost: During the Paleogene, Antarctica had not yet developed an ice cap. The continent may have had extensive permafrost and peat, which, when thawed, could have been the source of the carbon added during the PETM. A 2011 review article concluded that this was the most likely source of the carbon that brought on the PETM.[7]

Orbital cycles: The Late Paleocene and Early Eocene experienced several brief periods of global warming, of which the PETM was the most extreme. As we know by now, any repeating cycle of geological events raises the possibility that the orbital factors may have played a role. To test that idea, in May 2002 ODP Leg 208 drilled a section of the Atlantic Ocean floor off the southwest coast of Africa known as the Walvis Ridge. The core from site 1262 provided the most complete section across the P-E boundary available at the time. A 2010 article on the core reported that with one exception, the carbon isotope changes and climate at the hypothermal periods were paced by the precession and eccentricity cycles.

The one exception was the PETM, which seemed to be slightly out of phase with the orbital cycles.[8] However, a report from October 2022, based on ODP site 1262 and also on a geological section in Italy, found that the onset of the PETM occurred close to a maximum in the long eccentricity cycle.[9] On that basis, these authors concluded that the orbital cycles helped to bring on the PETM by triggering one of the other processes above.

Running Out of Time

For us, the critical information is how the amount of carbon added to the atmosphere compares with the amount that human activities are now releasing. Figure 16.1 shows how the amount of CO_2 in the atmosphere has increased steadily since measurements began in 1958.

Look carefully at the curve and you can see that it is concave upward, meaning that the rate of increase—how many parts per million of CO_2 human activities produce each year—has itself increased, getting larger by the year. This is no optical illusion: from 1959 to 2015, the annual *rate* of emissions quadrupled.[10] Today, human activities are emitting carbon at 10 times the rate of the PETM.

Biologist Philip Gingerich plotted the known rate of carbon emission through 2015 and extrapolated it into the future, as shown in figure 16.2.

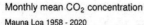

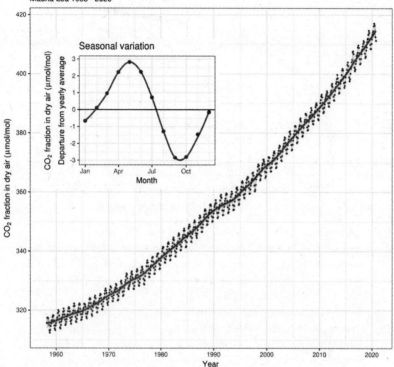

Figure 16.1

The Keeling Curve, named for Charles Keeling, who began the measurements. It shows that the amount of CO_2 in the atmosphere has risen from 316 ppm in 1959 to 416 ppm in 2021. Dots are the average monthly concentrations and the solid line is the smoothed trend. The inset shows the detail of the monthly variations, which reflect the seasonal removal of CO_2 by plants.

Source: Data from Pieter Tans and Ralph Keeling, Delorme, Wikimedia Commons, January 6, 2019, https://commons.wikimedia.org/wiki/Category:Keeling_Curves#/media/File:Mauna_Loa_CO2_monthly_mean_concentration.svg.

Time to PETM accumulation values

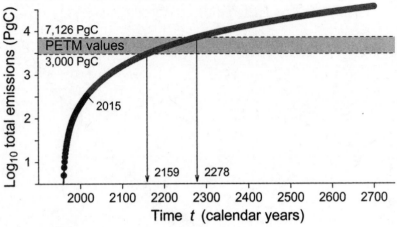

Figure 16.2
Projection of future carbon emissions if no action is taken to curtail
them. (PgC is petagrams of carbon, one of which is 10^{15} grams or one
gigaton. The vertical scale is logarithmic, so that 1, 2, 3, and 4 correspond to 10, 100, 1000, and 10,000.)
Source: Philip D. Gingerich, "Temporal Scaling of Carbon Emission and
Accumulation Rates: Modern Anthropogenic Emissions Compared to
Estimates of PETM Onset Accumulation," *Paleoceanography and Paleoclimatology* 34, no. 3 (2019): 333, https://doi.org/10.1029/2018PA003379.

If business-as-usual continues, Earth will reach the
minimum estimate for PETM-scale carbon emissions,
3,000 petagrams, in 2159, 136 years from now. By burning all known fossil fuel reserves, the planet would reach
the maximum for the P-E in the year 2278. To show that
even the minimum is not unimaginable, the emissions
trajectory the world is currently on, shown up to 2015
in the chart, is close to the business-as-usual emissions
scenario SSPS 8.5 of the IPCC. Unless we take strong

action to curtail global warming and get off that trajectory, those alive in the second half of the twenty-second century could experience conditions not seen since the Early Eocene.

To make that point real for us, Gingerich uses a poignant personal comparison that I adapt to my own history. The year 2159 lies 136 years from the time of this writing in 2023. My grandfather, Lawrence C. Powell, a Kentucky gentleman whom I remember well and with love, was born in 1884, 139 years ago. Thus, his lifetime and mine together span the time humanity has left before, on the present trajectory, PETM temperatures return. Or, to use a different comparison, consider that many of the great-grandchildren of people born in this decade will live to see 2159. Global warming does not lie in some far-off future or as some scenario of science fiction; it is coming and coming faster than we think, on a timetable measured not in millennia but in human generations. If global warming follows the path of the PETM, it will depart in 170,000 years. We can still prevent this catastrophic outcome, but to do so will require nations to forego their focus on the present, overcome their comparatively trivial partisan differences, recognize manmade global warming as the dire threat it is, and act for the sake of future generations—their own grandchildren.

17
Human Evolution

One of the key questions that scientists have long pondered is the extent to which climate change has affected evolution and, in particular, human evolution. In chapter III of *On the Origin of Species*, Charles Darwin wrote, "Each species . . . is constantly suffering enormous destruction at some period of its life, from enemies or from competitors for the same place and food; and if these enemies or competitors be in the least degree favoured by any slight change of climate, they will increase in numbers, and, as each area is already fully stocked with inhabitants, the other species will decrease."[1] At this stage in his writings, Darwin said little about human evolution other than to affirm that humans, like all other organisms, have evolved. However, in the introduction to *The Descent of Man*, published 12 years later, in which he first used the word evolution, he wrote:

> It has often and confidently been asserted, that man's
> origin can never be known: but ignorance more
> frequently begets confidence than does knowledge: it is
> those who know little, and not those who know much,

who so positively assert that this or that problem will
never be solved by science. The conclusion that man is
the co-descendant with other species of some ancient,
lower, and extinct form, is not in any degree new.[2]

Darwin did not write, as has commonly been claimed,
that man "descended from the apes." Rather, he proposed that they had a common ancestor.

Human Evolution and Climate Change

As scientists uncovered more fossil evidence of ancient
humans and as it became possible to date them accurately, primarily using the K–Ar method, it became clear
that there have been two major periods of change in
the evolutionary path of modern humans, separated by
roughly one million years.[3] The first occurred between
about 2.9 and 2.4 million years ago. Early in this period,
Australopithecus afarensis (whose most famous fossil was
named Lucy) went extinct and two new groups replaced
them. One represented the first appearance of *Homo*.
These fossils had the larger brains that have characterized
our genus ever since, perhaps the key to allowing them
to develop and create the stone tools of the Oldowan tradition found at Olduvai Gorge in Tanzania, the earliest
evidence of human cultural behavior. The second group
was quite different, with a heavy jaw and stout build.
This lineage, *Paranthropus*, also known as the robust aus-
tralopithecines, became extinct 1.2 million years ago.

The next period of great change took place between about 1.9 and 1.6 million years ago, with the arrival of *Homo erectus* (upright man). This ancestor had a more modern stride and displayed the flat face, large nose, teeth, and possibly the body hair, of modern humans. *H. erectus* produced the more advanced Acheulean stone tool culture, which included bifacial hand axes. The Neanderthals and modern humans are believed to have evolved from *H. erectus*. It also was the first species to migrate out of Africa to Europe and as far as Southeast Asia.

Hominin fossils—modern humans and our close extinct relatives—represent less than 1 percent of all African fossils, denying paleoanthropologists much of the evidence they need to assess the effects of climate on human evolution. For a more complete picture, they have turned to other mammals that occupied East Africa alongside our ancestors, but which have left many more fossils. One such is the bovid group: mammals of the cattle family that includes antelope, buffalo, sheep, goats, and so on. They make up about one-third of African fossils and their earliest and latest appearances in the fossil record have been well studied, notably by Professor Elisabeth Vrba of Yale University.[4] Instead of a steady state of evolutionary change, the first appearances of new bovid species cluster near 2.8 million and 1.8 million years ago, the same pattern as the rarer hominin fossils.

Darwin had pictured evolution as something like a steady-state process, proceeding at a roughly constant geological pace. But that view has long been seen to be too simplistic, as expressed in the punctuated equilibria model of paleontologists Stephen Jay Gould and Niles Eldredge. According to their theory, evolution is characterized by long periods of little or no change, interrupted by brief bursts of rapid species development. Scientists are coming to the view that climate change was the root cause of the two periods of relatively rapid evolutionary change described above, which had been preceded by a long period of steady change of the type Darwin envisioned. During the two periods, a broad shift was underway toward arid, open grasslands. But while this change was happening, on a smaller scale, the climate swung rapidly between wet and dry conditions, to which our ancestors had to adapt or go extinct.

This modern view has replaced a long-standing model known as the savanna hypothesis. Its key belief was that ancient hominins had lived primarily in trees, but descended to walk on two legs (bipedalism) on the savanna—grassy plains with few trees—that were replacing the woodlands.

South African paleoanthropologist Raymond Dart (1893–1988) made the critical discovery that led to the savanna hypothesis. Born in Australia, he studied medicine there and in 1922 became professor of anatomy at the University of the Witwatersrand in Johannesburg, South Africa. In 1924, a colleague sent him two boxes of fossils from the South African village of Taung. One specimen was a skull that Dart recognized had a brain cavity distinctly larger than those of a chimpanzee or baboon. He named and described it in a 1925 article titled *"Australopithecus africanus*: The Man-Ape of South Africa."[5] As is often the fate of the outsider, the European anthropology establishment rejected Dart's claim, viewing the Taung skull as that of an ape. But further research would vindicate him.

One byproduct of Dart's research was to highlight the role of climate in the evolution of early humans. In his 1925 article, he described what became known as the savanna hypothesis:

> For the production of man a different apprenticeship was needed to sharpen the wits and quicken the higher manifestations of intellect—a more open veldt country where competition was keener between swiftness and stealth, and where adroitness of thinking and movement played a preponderating role in the preservation of the species. . . . [I]n my opinion, Southern Africa, by providing a vast open country with occasional wooded belts and a relative scarcity of water, together with a fierce and bitter mammalian competition, furnished a laboratory such as was essential to this penultimate phase of human evolution.[6]

The savanna hypothesis was easily understood and found popular appeal in books such as those by Robert Ardrey, including his *The Territorial Imperative*.[7] He was a successful screenwriter and novelist who returned to his scientific roots, met Dart, and studied his specimens. But as more human fossils were found, they began to call the savanna hypothesis into question. Then came the deep-sea drilling cores which showed that instead of this climactic and evolutionary one-time shift, the East African climate had switched rapidly and frequently from warmer to cooler and back again. Moreover, scientists discovered that our ancestors walked upright millions of years before the grassy savannas expanded.[8] Though the savanna hypothesis was eventually disproven, it served a useful purpose.

From Dust to Dust

Before scientific ocean drilling, information about the geologic past of East Africa had to come from sediments collected on land, where rocks had been subject to erosion or been covered by more recently deposited sediments, destroying or hiding the evidence. What was needed was a continuous, undisturbed record—the very thing the deep-sea drilling cores provided. Two ODP sites have supplied critical information about African climate. In the early spring of 1986, Leg 108 of the ODP drilled 12 holes off the west coast of Africa from Mauritania to Liberia. These seafloor sediments record the supply of dust from the continent, which depends mainly on the extent of West African drought. When drought is prolonged, vegetative cover is reduced and dust storms become more frequent. Thus, the amount of windblown dust in marine sediments reveals the extent of aridity on nearby land surfaces. Leg 117 drilled in the fall of 1987 on the other side of Africa, off the coast of Oman in the Arabian Sea. These cores record information about the dust supply from northeast Africa as well as Arabia and Mesopotamia. The results of the dust measurements for seven sites from the two legs are shown in figure 17.1.

The first thing to strike the eye in figure 17.1 are the extremely rapid changes in both dust percentages and oxygen isotope ratios ($\delta^{18}O$). These are the exact

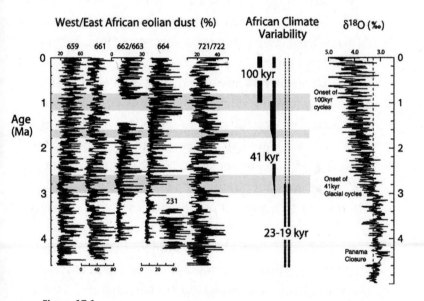

Figure 17.1
African windblown dust, climate, and oxygen isotope ratios from DSDP/ODP drilling sites.
Source: Peter B. deMenocal, "African Climate Change and Faunal Evolution during the Pliocene–Pleistocene," *Earth and Planetary Science Letters* 220, no. 1–2 (2004): 7.

opposite of the slow, steady change envisioned by the savanna hypothesis. By now, we have learned that such rapid but regular sedimentary fluctuations are due to the orbital cycles. Indeed, spectral analysis of the dust percentages, like that performed by Imbrie on the so-called Rosetta Stone core in the 1970s, confirms that conclusion. Before about 2.8 million years ago, the 23,000- and 19,000-year precessional cycles dominated; from then until about 1 million years ago, the 41,000-year obliquity cycle held sway; and from then until the present, the 100,000-year eccentricity cycle prevailed. These periods correspond roughly to the major changes in human evolution noted above.

In the Mediterranean region, East Africa included, anoxic, sapropel-producing conditions such as those described in chapter 12, have alternated repeatedly with ones in which abundant oxygen led to the removal of organic material, leaving those layers lighter in color. But what caused the alternation? The answer is the African monsoon. These wet and dry cycles occur because land and ocean water absorb and release heat differently. During summer months, sunlight heats both the land and the adjacent ocean, but because land absorbs more heat than does water, land temperatures rise faster and higher than those of the nearby ocean. The heat causes air above the land to expand, producing an area of low pressure. Moist air over the oceans, under higher pressure, flows inland to the low-pressure area where it rises, cools, and condenses to produce rain, sometimes in enormous volumes. In winter, the process reverses: The land cools faster than the ocean, creating higher pressure above the land surface. Air flows from the land to the sea, where it cools and causes rain to fall over the ocean. During the dry periods, monsoon winds blow dry dust over the region, which settles into dark layers. They are then covered by fine-grained, light-colored layers containing abundant carbonate shells of forams and other microfossils.

All this is revealed in the sediment cores. They show conclusively that the history of East African climate for the last four million years has been one of alternating wet and dry periods, not the single shift in climate as envisaged by the savanna hypothesis. Evidence from the most recent wet period showed that North Africa had gotten very wet indeed. From about 15,000 years ago to about 5,000, the precession cycle led to warmer summers and more rainfall in North Africa. The modern Sahara Desert was almost completely vegetated, with large, permanent lakes and abundant fauna. We know from ancient Saharan fossils and cave paintings that animals now found only in wetter parts of Africa—including antelope, elephants, giraffes, hippos, ostriches, and rhinos—roamed this "Green Sahara." The heavy rainfall produced greater runoff from the Nile, which dumped large amounts of fresh water into the eastern Mediterranean. As in the Black Sea today, the fresh water sat on top, turning the deeper water anoxic, where one sapropel layer after another accumulated as the precession cycle waxed and waned. The enhanced monsoon began to ebb about 6,500 years ago, and by about 4,000 years ago, the Sahara had taken on the extreme aridity by which we know it today.[9]

In 1974, Leg 24 of the DSDP drilled Hole 231 in the Gulf of Aden, 80 km off the coast of what is today recognized as Somaliland. It provides another example of the wisdom of creating core repositories, because 39 years later this core was used for a study no one could have conceived of at the time it was collected. The method depended on a waxy substance that coats the leaves of terrestrial plants, helping them to retain water and protecting them from injury. These waxlike coatings last for a long time and can be chemically extracted from sediment cores and analyzed for their carbon isotope ratios, which vary between grasslands and woodlands. In 2013, Sarah Feakins of the University of Southern California and her students and colleagues reported that the waxy leaf coatings from Hole 231 showed that East African grasslands had grown larger by as much as 50 percent between three and two million years ago.[10] But they also found the much shorter back and forth alternations due to the orbital cycles, confirming the conclusions from the dust studies. Other scientists used the same technique on East African lake sediments and found the same repeated ecosystem swings timed by precession.

Survival of the Adaptable

Despite the East African climate having switched from wet to dry and back again on the ~21,000 year time-scale of the precession cycles, hominin and bovid species managed to survive for hundreds of thousands of years. Rick Potts, director of the Human Origins Program at the Smithsonian Institution, used this fact to come up with a novel theory of human evolution that he nick-named "survival of the adaptable."[11] The point of his theory was that an important selection factor, if not the most important, in the evolution of African hominins may have been their ability to adapt to extreme climate variability itself. Potts and paleoanthropologist J. Tyler Faith were able to test this hypothesis by predicting when the orbital cycles would cause periods of high and low climate variability.[12] They compared those predic-tions with deep-sea drilling records of northeast African dust flux and eastern Mediterranean sapropels cored at ODP Site 967A, finding that the timetable of changes in environmental records from hominin fossil–bearing basins in eastern Africa matched the predicted timing of variability changes. Potts wrote, "A species depends not only on the particular way of life that best matches its surroundings but also on keeping certain options open and adjusting to whatever trials or opportunities occur as things change."[13] The ability to walk upright and cover long distances may have given *Homo erectus* the ability to migrate and survive in the new environments brought on by climate change. Later they would have enabled the exodus from Africa. The stone toolkit developed by early hominins would have allowed them to prepare and eat new kinds of food as the climate began to disfavor their traditional diet. Managing fire, erecting crude shelters, and growing and storing food—the "entire package" as Potts writes—"proved so successful that, eventually, the sole surviving hominin—*Homo sapiens*—was able to spread around the globe."

18
Looking Back and Ahead

During the twentieth century, scientific ocean drilling brought more advances in earth science than did any research program. Some were truly revolutionary. Each of the questions with which we began this book was answered, transforming the science of geology. Many of these advances could not have been made without scientific ocean drilling.

The members of the American Miscellaneous Society and National Science Foundation review panels, in their conference rooms of the late 1950s, could never have imagined where scientific ocean drilling would take geology. They were looking for a moon shot equivalent, a one-off that would draw the same kind of funding and attention to the earth sciences that other "big science" and space projects were receiving. Instead, the demise of Project Mohole opened the way for a vastly more productive program.

At the time the Mohole planners met, continental drift had been rejected for half a century, mainly because it violated the allegedly inviolable dogma of uniformitarianism. Yet it took only the third leg of the DSDP to confirm the paleomagnetic timescale and sea-floor spreading, opening the way to continental drift and the plate tectonics revolution.

Their allegiance to uniformitarianism also caused most midcentury scientists to reject meteorite impact cratering on the Moon and Earth. Even as late as 2010, many continued to oppose the then 30-year-old Alvarez Theory of dinosaur extinction by meteorite impact. But the results of drilling the Chicxulub crater in 2016 removed any remaining doubt that the theory is correct and showed that the peak rings seen in lunar craters came from deep below the surface, rather than from the collapse of the surficial central peaks.

In the 1950s, human-caused global warming was hardly recognized even as a possibility. Most scientists had never heard of the concept. Since then, the evidence has mounted and become irrefutable: Earth is undergoing dangerous warming, and it is our fault. The lack of action indicates that even some who accept that finding believe we can delay responding while we focus on the always-present, more urgent problems. Few have realized, or admitted to themselves, just how fast the danger is approaching. Cores from the hot, Early Eocene show that unless we sharply reduce carbon emissions, it is not some distant generation that will suffer, but our own great-grandchildren.

Another seemingly inexplicable scientific problem was the origin of the many island chains of the Pacific Ocean. James Dwight Dana thought that volcanism moving along a fissure in the upper crust caused them. J. Tuzo Wilson posited instead that the crust had moved over a stationary hotspot in the mantle. Drilling to measure the ages and paleolatitudes of the Hawai'ian-Emperor and Louisville chains showed that hotspots exist and move independently of each other. Indeed, plates and hotspots both move. This finding makes the picture more complex but provides important clues as to how the mantle operates, which in turn controls events on the surface.

The Milankovitch theory that astronomical cycles pace Earth's climate and caused the ice ages had risen and fallen in favor, never quite achieving consensus. But by the 1950s, a raft of new discoveries and instruments had become available to attempt to resolve the old controversy. These included the paleomagnetic timescale, potassium–argon dating, oxygen isotope paleothermometers, and especially, the deep-sea drilling cores. In 1976, Hays, Imbrie, and Shackleton used the Rosetta Stone core to confirm the Milankovitch theory. This in turn led to two new fields of science: astrochronology and cyclostratigraphy. Evidence of cyclostratigraphy has even turned up on Mars, confirming the dominance of the astronomical cycles and the one-time presence of water on the Red Planet. Even human evolution has had to answer to Earth's astronomical pacemaker.

Before scientific ocean drilling, the geological history of Antarctica, hidden by its ice cover, was a mystery. The continent is so large that until more of its history was known—for instance, when Antarctica glaciated—the history of the Southern Ocean also remained unknown.

A single leg of the DSDP provided enough information to allow a detailed account of Antarctica's geological evolution since about 65 million years ago. Not long ago in geological time, Antarctica was ice-free and the global sea level was much higher. Dangerous melting of Antarctic ice could happen again, this time on a human timescale.

One of the most surprising findings from scientific ocean exploration is that wherever scientists have looked beneath the seafloor, they have found microbial life. This could mean that microbes were present under extreme conditions elsewhere, such as the surface of Mars. Microbial life could still persist in the subsurface of the Red Planet. Life could have begun with bacteria that used chemosynthesis first, then evolved to photosynthesis.

One important byproduct of scientific ocean drilling was to diversify and internationalize geology—to "democratize" the science, as geophysicist and former president of the National Academy of Sciences Marcia McNutt has put it.[1] AMSOC had no female members and the typical NSF review panel of the 1950s had few, if any. Nor for that matter, did most university geology faculties. All that began to change with the women's movement of the 1960s. Women have played an increasingly important role in the geosciences and in scientific ocean drilling ever since, as a glance at any recent IODP crew photo will show. Many have served as crew chief.

Scientists from 24 countries have participated in the IODP, with the United States and Japan being the most represented. The leading vessel in the IODP is the

Japanese *Chikyū*, but after an overhaul, the indefatigable *JOIDES Resolution* is still at work. These diverse scientists have collaborated in the more than 26,000 research publications that have resulted from four decades of deep-sea drilling.[2]

The IODP is scheduled to end on September 30, 2024. Efforts have been underway since 2020 to plan the "2050 Science Framework: Exploring Earth by Scientific Ocean Drilling."[3] Scientists and funders are conducting a series of workshops to design the new drillship and set out the objectives of this next phase of the revolutionary research program. The European Consortium for Ocean Research Drilling, which includes 15 countries, will play an integral role. Planning workshops have engaged more than 650 scientists from many nations, with significant representation by women (30 percent) and early-career scholars (40 percent).

The planners have identified five flagship initiatives for the new program:

- Ground-truthing future climate change
- Probing the deep earth
- Diagnosing ocean health
- Exploring life and its origins
- Assessing earthquake and tsunami hazards

The last of these deserves special mention. Deep-sea drilling can shed light on what scientists refer to as geohazards: geologic events that can cause the destruction of significant life and property. These include submarine earthquakes that can cause damage directly and through the triggering of tsunami. The deadly December 26, 2004 Indian Ocean tsunami resulted from an earthquake caused by the movement of two tectonic plates. It was the third largest earthquake ever recorded, and the tsunami it generated killed nearly 230,000 people. On March 11, 2011, off the east coast of Honshu, there occurred the most powerful earthquake ever recorded in Japan and the fourth largest anywhere in the world since modern recordkeeping began. It gave rise to tsunami waves up to 40 m high and traveling at 700 km/h. Nearly 20,000 people died from the tsunami. It also swamped an insufficiently high seawall at the Fukushima Daiichi Nuclear Power Plant and flooded the generators that were needed to cool the reactor, causing the most severe nuclear accident since the 1986 Chernobyl disaster. Other geohazards include gravity-induced or earthquake-triggered debris flows and volcanic eruptions that send enormous amounts of ash, steam, and volcanic ejecta into the atmosphere. So far in scientific ocean drilling, the study of these and other geohazards for the most part has been incidental to other drilling studies. But in recognition of their importance, a directed geohazards component will constitute a major thrust of the 2050 Science Framework.

Vannevar Bush served as dean of the MIT School of Engineering in the 1930s. He went on to direct the US Office of Scientific Research and Development during World War II and become the first presidential science advisor. His 1945 report to the president, titled *Science:*

Eon	Era/Subera	Period/Subperiod		Epoch	Age (million years ago)
Phanerozoic	Cenozoic (Tertiary)	Quaternary		Holocene	0.01
				Pleistocene	1.6
		Neogene		Pliocene	5.2
				Miocene	23
		Paleogene		Oligocene	35
				Eocene	56
				Paleocene	65
	Mesozoic	Cretaceous			146
		Jurassic			208
		Triassic			248
	Paleozoic	Permian			290
		Carboniferous	Pennsylvanian		323
			Mississippian		362
		Devonian			408
		Silurian			439
		Ordovician			510
		Cambrian			570
Precambrian	Proterozoic				2500
	Archean				4000
	Priscoan				4490

Figure 18.1
The geologic time scale.

The Endless Frontier, led to the establishment of the National Science Foundation.[4] Bush wrote, "It has been basic United States policy that Government should foster the opening of new frontiers. It opened the seas to clipper ships and furnished land for pioneers. Although these frontiers have more or less disappeared, the frontier of science remains." Never was this more clear than in the story of scientific ocean drilling, now in its fifty-fifth year, with an endless frontier ahead.

Appendix:
Vessels and Programs

The original Deep Sea Drilling Project (DSDP) operated from 1968 to 1983 under the direction of the Scripps Institute of Oceanography and with funding from the National Science Foundation. It used the research vessel *Glomar Challenger*, built by Global Marine, Inc. An advisory group named Joint Oceanographic Institutions for Deep Earth Sampling (JOIDES) planned the operations. It comprised some 250 scientists from academic institutions, government agencies, and private industry from many nations, establishing what was a truly international program.

In 1985, the new Ocean Drilling Program (ODP) began, using a new vessel, the *JOIDES Resolution*, named in honor of the flagship of famed explorer Captain James Cook. It was originally an oil exploration ship but had been converted for scientific research. In 2007–2008 *Resolution* was renovated and modernized. At the establishment of ODP, JOIDES had nearly two dozen universities and international science organizations as members. JOIDES served as ODP program manager and Texas A&M University as science operator.

The ODP was followed by the Integrated Ocean Drilling Program (IODP) beginning in 2003, still using the *JOIDES Resolution*. Now instead of using the term "legs," the IODP used "expeditions." In 2013, it was renamed the International Ocean Discovery Program, with the same acronym. The *JOIDES Resolution* was now joined by a Japanese-built drillship, the *Chikyū*, which means "Earth." It was designed to drill 7 km through the seafloor crust and into Earth's mantle, the original goal of the abandoned Project Mohole. Scientists planning the 2050 Science Framework are designing a new drillship.

Acknowledgments

With thanks to my longtime agent John Thornton, to Jonathan Cobb for expert editing, and to Carol Shetler for proofreading. My deepest thanks to my former student, the late Peter Molnar, for his careful reading of the manuscript.

Notes

Chapter 1

1. Stephen Jay Gould, "Is Uniformitarianism Necessary?," *American Journal of Science* 263, no. 3 (1965): 223–228.

Chapter 2

1. C. W. Thomson, *Report on the Scientific Results of the Voyage of H.M.S.* Challenger *during the Years 1873–76, Under the Command of Captain George S. Nares, R.N., F.R.S. and the Late Captain Frank Tourle Thomson, R.N.*, vol. 1 (London: FB&C, 1885), li–lii, http://www.19thcenturyscience.org/HMSC/HMSC-Reports/1885-Narrative/htm/doc.html.

2. Richard Corfield, *The Silent Landscape: In the Wake of HMS* Challenger, *1872–1876* (London: John Murray, 2004), Kindle, 4–5.

3. George Campbell, *Log-Letters from "The* Challenger*"* (London: Macmillan, 1877), 4–5, https://books.google.com/books?id=l7IEAAAAYAAJ.

4. Thomson, *Report on the Scientific Results of the Voyage of H.M.S.* Challenger, http://www.19thcenturyscience.org/HMSC/HMSC-INDEX/index-illustrated.htm.

5. Thomas R. Anderson and Tony Rice, "Deserts on the Sea Floor: Edward Forbes and His Azoic Hypothesis for a Lifeless Deep Ocean," *Endeavour* 30, no. 4 (December 2006): 131–137, https://doi.org/10.1016/j.endeavour.2006.10.003.

6. Kate Golembiewski, "H.M.S. *Challenger*: Humanity's First Real Glimpse of the Deep Oceans," *Discover Magazine*, April 19, 2019, https://www.discovermagazine.com/planet-earth/hms-challenger -humanitys-first-real-glimpse-of-the-deep-oceans.

7. Louis Agassiz, *Outlines of Comparative Physiology* (London: H. G. Bohn, 1850), 365, https://archive.org/details/in.ernet.dli.2015.221191 /page/n3/mode/2up?q=impassable.

8. James Lawrence Powell, *Four Revolutions in the Earth Sciences: From Heresy to Truth* (New York: Columbia University Press, 2014); James Lawrence Powell, "Premature Rejection in Science: The Case of the Younger Dryas Impact Hypothesis," *Science Progress* 105, no. 1 (January–March 2022), https://doi.org/10.1177/00368504211064272.

9. T. H. Huxley, "De Partibus Similaribus," *British and Foreign Medico-Chirurgical Review* 12, no. 24 (October 1853): 291.

10. T. H. Huxley, "The Problems of the Deep Sea," in *Popular Science Monthly*, vol. 3 (New York: D. Appleton, 1873): 15.

11. Elizabeth Cary Agassiz, "Chapter 23: 1871–1872: Aet. 64–65," in *Louis Agassiz: His Life and Correspondence*, 3rd ed. (New York: Houghton Mifflin, 1885), 708, https://www.perseus.tufts.edu/hopper/text ?doc=Perseus%3Atext%3A2001.05.0258%3Achapter%3D24.

12. Loren Eiseley, *The Immense Journey: An Imaginative Naturalist Explores the Mysteries of Man and Nature* (New York: Knopf Doubleday, 2011), 41.

13. Henry Nottidge Moseley, *Notes by a Naturalist on the* Challenger (Cambridge, UK: Cambridge University Press, 2014), 586.

14. Thomson, *Report on the Scientific Results of the Voyage of H.M.S. Challenger*, 194.

15. Corfield, *The Silent Landscape*, loc. 2013.

16. Corfield, loc. 956, 969.

17. Maya Wei-Haas, "Accidental Implosion Yields New Measurement for Ocean's Deepest Point," *National Geographic*, February 8, 2022, https://www.nationalgeographic.com/science/article/accidental -implosion-yields-new-measurement-for-oceans-deepest-point.

18. Lyndsey Fox et al., "Quantifying the Effect of Anthropogenic Climate Change on Calcifying Plankton," *Scientific Reports* 10 (2020): 1620, https://doi.org/10.1038/s41598-020-58501-w.

Chapter 3

1. Gordon Chancellor, "Introduction to Coral Reefs," Darwin Online, accessed February 18, 2022, http://darwin-online.org.uk/Editorial Introductions/Chancellor_CoralReefs.html.

2. Sandra Herbert, *Charles Darwin, Geologist* (Ithaca, NY: Cornell University Press, 2005), 168.

3. Charles Darwin, *The Structure and Distribution of Coral Reefs. Being the First Part of the Geology of the Voyage of the Beagle, under the Command of Capt. Fitzroy, R.N. during the Years 1832 to 1836* (London: Smith, Elder, 1842).

4. Darwin, 110.

5. Darwin, 227.

6. Chancellor, "Introduction to Coral Reefs."

7. John Murray, "1. On the Structure and Origin of Coral Reefs and Islands," *Proceedings of the Royal Society of Edinburgh* 10 (1880): 505–518.

8. Iain McCalman, *The Reef: A Passionate History: The Great Barrier Reef from Captain Cook to Climate Change* (New York: Farrar, Straus and Giroux, 2014), 191.

9. Charles Darwin, *The Life and Letters of Charles Darwin: Including an Autobiographical Chapter* (London: John Murray, 1888), 184.

10. Smithsonian Institution Board of Regents, *Annual Report of the Board of Regents of the Smithsonian Institution* (Washington, DC: The Institution, 1899), 406.

11. H. S. Ladd, J. I. Tracey, and G. G. Lill, "Drilling on Bikini Atoll, Marshall Islands," *Science* 107, no. 2768 (January 16, 1948): 51–55, https://doi.org/10.1126/science.107.2768.51.

12. Harry Hammond Hess, "Drowned Ancient Islands of the Pacific Basin," *American Journal of Science* 244, no. 11 (1946): 772–791, https://doi.org/10.2475/ajs.244.11.772.

Chapter 4

1. Penelope Hardy, "Meteorology as Nationalism on the German Atlantic Expedition, 1925–1927," *History of Meteorology* 8, no. 8 (2017): 124.

2. Gøsta Walin and Ingemar Olsson, "Professor Börje Kullenberg 1906–1991," *ICES Journal of Marine Science* 50, no. 1 (January 1, 1993): 101–102, https://doi.org/10.1006/jmsc.1993.1011.

3. Edward Bullard, "William Maurice Ewing," *Biographical Memoirs* 51 (1980): 124, https://nap.nationalacademies.org/read/574/chapter/8. The following paragraphs draw heavily from this source.

4. Bullard, 160.

5. David M. Lawrence, *Upheaval from the Abyss: Ocean Floor Mapping and the Earth Science Revolution* (New Brunswick, NJ: Rutgers University Press, 2002).

6. "History of the Core Repository | Lamont-Doherty Earth Observatory," accessed March 3, 2021, https://corerepository.ldeo.columbia.edu/content/history-core-repository.

7. Bullard, "William Maurice Ewing," 149.

8. For more on the Lamont-Doherty Core Repository at Columbia University, see https://corerepository.ldeo.columbia.edu.

9. William W. Hay and Eloise Zakevich, "Cesare Emiliani (1922–1995): The Founder of Paleoceanography," *International Microbiology* 2, no. 1 (1999): 52–54.

10. Cesare Emiliani, "Pleistocene Temperatures," *Geology* 63, no. 6 (November 1955): 538–578, https://doi.org/10.1086/626295.

11. National Science Foundation Act of 1950, H.R. 4346, 117th Cong. (2022).

12. John Steinbeck, "High Drama of Bold Thrust through Ocean Floor: Earth's Second Layer Is Tapped in Prelude to Mohole," *Life*, April 14, 1961.

13. Keir Becker et al., "50 Years of Scientific Ocean Drilling," *Oceanography* 32, no. 1 (2019): 17–18.

14. Steinbeck, "High Drama of Bold Thrust through Ocean Floor," 122.

15. D. S. Greenberg, "Mohole: The Project That Went Awry (II)," *Science* 143, no. 3603 (January 17, 1964): 224, https://www.science.org /doi/10.1126/science.143.3603.223.

16. Willard Bascom, "The Mohole Project," *Science* 226, no. 4671 (October 12, 1984): 114.

17. D. S. Greenberg, "Mohole: The Project That Went Awry (III)," *Science* 143, no. 3604 (January 24, 1964): 335, https://doi.org/10.1126 /science.143.3604.334.

Chapter 5

1. Alfred Wegener, "Die Entstehung der Kontinente," *Geologische Rundschau* 3, no. 4 (1912): 276–292; Alfred Wegener, *Die Entstehung der Kontinente und Ozeane* (Braunschweig: F. Vieweg & Sohn Akt.-Ges., 1922), https://books.google.com/books?id=AI2Cj67SIx8C.

2. Mott T. Greene, *Alfred Wegener: Science, Exploration, and the Theory of Continental Drift* (Baltimore, MD: Johns Hopkins University Press, 2015).

3. Marie Tharp, "Marie Tharp: Taken From 'Connect the Dots: Mapping the Seafloor and Discovering the Mid-Ocean Ridge' by Marie Tharp, chapter 2 in *Lamont-Doherty Earth Observatory of Columbia: Twelve Perspectives on the First Fifty Years 1949–1999*, edited by Laurence Lippsett," Woods Hole Oceanographic Institution, April 1, 1999, https://www .whoi.edu/news-insights/content/marie-tharp/. The information in this section comes mainly from this source and *The Conversation* article cited in the following note.

4. Suzanne OConnell, "Marie Tharp Pioneered Mapping the Bottom of the Ocean 6 Decades Ago—Scientists Are Still Learning about Earth's Last Frontier," *The Conversation*, updated February 28, 2022, https://theconversation.com/marie-tharp-pioneered-mapping-the -bottom-of-the-ocean-6-decades-ago-scientists-are-still-learning -about-earths-last-frontier-142451.

Chapter 6

1. Arthur Holmes, "XVIII. Radioactivity and Earth Movements," *Transactions of the Geological Society of Glasgow* 18, no. 3 (1931): 559–606.

2. F. J. Vine and D. H. Matthews, "Magnetic Anomalies Over Oceanic Ridges," *Nature* 199, no. 4897 (1963): 947–949.

Chapter 7

1. John Steinbeck, "High Drama of Bold Thrust through Ocean Floor: Earth's Second Layer Is Tapped in Prelude to Mohole," *Life*, April 14, 1961, 111.

2. Edward L. Winterer, "Scientific Ocean Drilling, from AMSOC to COMPOST," in *50 Years of Ocean Discovery: National Science Foundation, 1950–2000* (Washington, DC: National Academies Press, 2000), 120, https://nap.nationalacademies.org/read/9702/chapter/13.

3. M. Ewing, J. L. Worzel, and C. A. Burk, "Introduction," in *Initial Reports of the Deep Sea Drilling Project*, vol. 1 (Washington, DC: US Government Printing Office, 1969), 3, http://deepseadrilling.org/01 /volume/intro.pdf.

4. Ewing, Worzel, and Burk, "Introduction," 6–9.

5. K. J. Hsu, Challenger *at Sea: A Ship That Revolutionized Earth Science* (Princeton, NJ: Princeton University Press, 1992).

6. Hsu, 88.

7. Hsu, 94.

8. George W. Gray, "The Lamont Geological Observatory," *Scientific American* 195, no. 6 (December 1956): 86–87.

9. Ellen S. Kappel and John W. Farrell, eds., "ODP's Greatest Hits" (Washington, DC: Joint Oceanographic Institutions, November 1997), http://www.odplegacy.org/PDF/Outreach/Brochures/ODP_Great est_Hits.pdf.

10. EarthByte, "Plate Tectonic Evolution from 1 Billion Years Ago to the Present," YouTube, February 1, 2021, video, 0:40, https://youtu .be/gQqQhZp4uG8.

11. J. R. Heirtzler et al., "Marine Magnetic Anomalies, Geomagnetic Field Reversals, and Motions of the Ocean Floor and Continents," *Journal of Geophysical Research (1896–1977)* 73, no. 6 (1968): 2119–2136, https://doi.org/10.1029/JB073i006p02119.

12. J. Tuzo Wilson, "A New Class of Faults and Their Bearing on Continental Drift," *Nature* 207, no. 4995 (July 1965): 343–347, https://doi .org/10.1038/207343a0.

Chapter 8

1. Thomas S. Kuhn and Ian Hacking, *The Structure of Scientific Revolutions: 50th Anniversary Edition*, 4th ed. (Chicago: University of Chicago Press, 2012).

2. James D. Dana, "ART. XIII.—On the Origin of Continents," *American Journal of Science and Arts (1820–1879)* 3, no. 7 (1847): 94.

3. J. Tuzo Wilson, "Some Consequences of Expansion of the Earth," *Nature* 185, no. 4717 (March 1960): 880–882, https://doi.org/10.1038 /185880a0.

4. Gordon F. West et al., "John Tuzo Wilson: A Man Who Moved Mountains," *Canadian Journal of Earth Sciences* 51, no. 3 (March 2014): xvii–xxxi, https://doi.org/10.1139/cjes-2013-0175.

5. Robert S. Dietz, "Continent and Ocean Basin Evolution by Spreading of the Sea Floor," *Nature* 190, no. 4779 (1961): 854–857.

6. Robert S. Dietz, "Earth, Sea, and Sky: Life and Times of a Journeyman Geologist," *Annual Review of Earth and Planetary Sciences* 22 (1994): 1–32.

7. J. Tuzo Wilson, "Mid-Ocean Ridges," *Nature* 192 (1961).

8. Alan Ott Allwardt, "The Roles of Arthur Holmes and Harry Hess in the Development of Modern Global Tectonics" (PhD diss., University of California, Santa Cruz, 1990), 190.

9. J. Tuzo Wilson, "Evidence from Islands on the Spreading of Ocean Floors," *Nature* 197 (1963): 536.

10. J. Tuzo Wilson, "A Possible Origin of the Hawaiian Islands," *Canadian Journal of Physics* 41, no. 6 (1963): 863–870, https://doi.org /10.1139/p63-094.

11. Wilson, "A Possible Origin of the Hawaiian Islands," 865.

12. J. Tuzo Wilson, "Evidence from Ocean Islands Suggesting Movement in the Earth," *Philosophical Transactions of the Royal Society of London. Series A, Mathematical and Physical Sciences* 258, no. 1088 (October 1965): 158, https://doi.org/10.1098/rsta.1965.0029.

13. W. Jason Morgan, "Convection Plumes in the Lower Mantle," *Nature* 230, no. 5288 (1971): 42–43.

14. Peter Molnar and Tanya Atwater, "Relative Motion of Hot Spots in the Mantle," *Nature* 246, no. 5431 (November 1973): 288–291, https://doi.org/10.1038/246288a0.

15. Masaru Kono, "Paleomagnetism of DSDP Leg 55 Basalts and Implications for the Tectonics of the Pacific Plate," in *Initial Reports of the Deep Sea Drilling Project,* vol. 55 (Washington, DC: US Government Printing Office, 1980), 737–752.

16. John Tarduno et al., "The Bent Hawaiian-Emperor Hotspot Track: Inheriting the Mantle Wind," *Science* 324, no. 5923 (April 3, 2009): 50–53, https://doi.org/10.1126/science.1161256.

Chapter 9

1. W. Jacquelyne Kious and Robert I. Tilling, "Historical Perspective," in *This Dynamic Earth: The Story of Plate Tectonics* (Washington, DC: US Government Printing Office, 1996), https://pubs.usgs.gov/gip /dynamic/historical.html.

2. Pliny, *Natural History,* vol. 2: books 3–7, trans. H Rackham (Cambridge, MA: Harvard University Press, 1942).

3. W. B. Ryan et al., *Initial Reports of the Deep Sea Drilling Project*, vol. 13 (Washington, DC: US Government Printing Office, 1973).

4. K. J. Hsu, Challenger *at Sea: A Ship That Revolutionized Earth Science* (Princeton, NJ: Princeton University Press, 1992); William Ryan and Walter Pitman, *Noah's Flood: The New Scientific Discoveries about the Event that Changed History* (New York: Simon & Schuster, 2000).

5. Ryan and Pitman, *Noah's Flood*, 73.

6. Hsu, Challenger *at Sea,* 257.

7. This section is drawn mainly from Hsu, Challenger *at Sea*, and Ryan and Pitman, *Noah's Flood*.

8. Hsu, Challenger *at Sea*, 11.

9. Ryan and Pitman, *Noah's Flood*, 86.

10. K. J. Hsu, M. B. Cita, and W. B. F. Ryan, "The Origin of the Mediterranean Evaporites," in *Initial Reports of the Deep Sea Drilling Project*, vol. 13 (Washington, DC: US Government Printing Office, 1973), 1203.

11. Ryan and Pitman, *Noah's Flood*, 85.

12. Daniel Garcia-Castellanos et al., "The Zanclean Megaflood of the Mediterranean—Searching for Independent Evidence," *Earth-Science Reviews* 201 (February 2020), https://doi.org/10.1016/j.earscirev.2019.103061.

Chapter 10

1. Charles Darwin, *Autobiography and Selected Letters* (Mineola, NY: Dover Publications, 1958), 26.

2. Darwin.

3. John Imbrie and Katherine Palmer Imbrie, *Ice Ages: Solving the Mystery* (Cambridge, MA: Harvard University Press, 1986). Much of the material in this chapter is drawn from this fine account of the discovery of the cause of the ice ages by these father and daughter coauthors. Quotations without references in this chapter come from this book.

4. James Lawrence Powell, *Night Comes to the Cretaceous: Dinosaur Extinction and the Transformation of Modern Geology* (New York: W. H. Freeman, 1998).

5. Andrei G. Lapenis, "Arrhenius and the Intergovernmental Panel on Climate Change," *Eos, Transactions, American Geophysical Union* 79, no. 23 (June 1998): 269–271, https://doi.org/10.1029/98EO00206.

6. "Climate Sensitivity," MIT Climate Portal, last updated June 8, 2023, https://climate.mit.edu/explainers/climate-sensitivity.

7. James Lawrence Powell, *Four Revolutions in the Earth Sciences: From Heresy to Truth* (New York: Columbia University Press, 2014).

8. Editors of Encyclopaedia Britannica, "Polestar," *Britannica*, last updated March 11, 2013, https://www.britannica.com/science/polestar.

9. Ronald Grimsley, "Jean Le Rond d'Alembert," *Britannica*, last updated November 13, 2022, https://www.britannica.com/biography/Jean-Le-Rond-dAlembert.

10. Alan Buis, "Why Milankovitch (Orbital) Cycles Can't Explain Earth's Current Warming," Global Climate Change: Vital Signs of the Planet, February 27, 2020, https://climate.nasa.gov/ask-nasa-climate/2949/why-milankovitch-orbital-cycles-cant-explain-earths-current-warming/.

11. James Croll, *Climate and Time in Their Geological Relations: A Theory of Secular Changes of the Earth's Climate* (New York: D. Appleton, 1875).

Chapter 11

1. John Imbrie and Katherine Palmer Imbrie, *Ice Ages: Solving the Mystery* (Cambridge, MA: Harvard University Press, 1986). Much of the material in this chapter is drawn from this fine account of the discovery of the cause of the ice ages by these father and daughter coauthors. Quotations without references in this chapter come from this book.

2. Milutin Milankovitch, *Mathematical Theory of Heat Phenomena Produced by Solar Radiation* (Paris: Gauthier-Villaras, 1920).

3. Kenneth J. Mesolella et al., "The Astronomical Theory of Climatic Change: Barbados Data," *Journal of Geology* 77, no. 3 (May 1969): 250–274, https://doi.org/10.1086/627434.

4. N. D. Opdyke et al., "Paleomagnetic Study of Antarctic Deep-Sea Cores," *Science* 154, no. 3747 (1966): 349–357.

5. An inscribed monument or "stele" discovered in Egypt in 1799. It contained the same text in hieroglyphics, Demotic, and Greek, and was key to the decipherment of Egyptian writing.

6. J. D. Hays, John Imbrie, and N. J. Shackleton, "Variations in the Earth's Orbit: Pacemaker of the Ice Ages," *Science* 194, no. 4270 (1976): 1121–1132.

Chapter 12

1. John Imbrie et al., "The Orbital Theory of Pleistocene Climate: Support from a Revised Chronology of the Marine d18O Record," in *Milankovitch and Climate, Part 1*, ed. A. L. Berger et al. (Dordrecht: D. Reidel, 1984), 269–305.

2. N. J. Shackleton, A. Berger, and W. R. Peltier, "An Alternative Astronomical Calibration of the Lower Pleistocene Timescale Based on ODP Site 677," *Earth and Environmental Science Transactions of the Royal Society of Edinburgh* 81, no. 4 (1990): 251–261, https://doi.org/10.1017/S0263593300020782.

3. A. K. Baksi et al., "40Ar/39Ar Dating of the Brunhes-Matuyama Geomagnetic Field Reversal," *Science* 256, no. 5055 (1992): 356–357, doi: 10.1126/science.256.5055.356.

4. F. J. Hilgen, "Extension of the Astronomically Calibrated (Polarity) Time Scale to the Miocene/Pliocene Boundary," *Earth and Planetary Science Letters* 107, no. 2 (November 1991): 349–368, https://doi.org/10.1016/0012-821X(91)90082-S.

5. Kate Littler et al., "Astronomical Time Keeping of Earth History," in "Scientific Ocean Drilling: Looking to the Future," special issue, *Oceanography* 32, no. 1 (March 2019): 72–76.

6. Thomas Westerhold et al., "An Astronomically Dated Record of Earth's Climate and Its Predictability over the Last 66 Million Years," *Science* 369, no. 6509 (September 11, 2020): 1383–1387, https://doi.org/10.1126/science.aba6853.

7. Linda A. Hinnov, "Astronomical Metronome of Geological Consequence," *Proceedings of the National Academy of Sciences* 115, no. 24 (May 2018): 6104–6106, https://doi.org/10.1073/pnas.1807020115.

8. Grove Karl Gilbert, "Sedimentary Measurement of Cretaceous Time," *Journal of Geology* 3, no. 2 (February–March 1895): 121–127, https://www.jstor.org/stable/30054556.

9. Robert E. Locklair and Bradley B. Sageman, "Cyclostratigraphy of the Upper Cretaceous Niobrara Formation, Western Interior, USA: A Coniacian–Santonian Orbital Timescale," *Earth and Planetary Science Letters* 269, no. 3–4 (2008): 540–553.

10. Alfred G. Fischer, "Climatic Rhythms Recorded in Strata," *Annual Review of Earth and Planetary Sciences* 14 (May 1986): 351–376.

11. John W. Wells, "Coral Growth and Geochronometry," *Nature* 197, no. 4871 (March 1963): 948–950, https://doi.org/10.1038/197948a0.

12. Aske L. Sørensen et al., "Astronomically Forced Climate Change in the Late Cambrian," *Earth and Planetary Science Letters* 548 (October 15, 2020), https://doi.org/10.1016/j.epsl.2020.116475.

Chapter 13

1. James Lawrence Powell, *Night Comes to the Cretaceous: Dinosaur Extinction and the Transformation of Modern Geology* (New York: W. H. Freeman, 1998).

2. Glenn L. Jepsen, "Riddles of the Terrible Lizards," *American Scientist* 52, no. 2 (June 1964): 227–246.

3. Luis W. Alvarez et al., "Extraterrestrial Cause for the Cretaceous-Tertiary Extinction," *Science* 208, no. 4448 (June 1980): 1095–1108.

4. Powell, *Night Comes to the Cretaceous*, 159; 216–217.

5. Powell, 103.

6. Walter Alvarez et al., "Proximal Impact Deposits at the Cretaceous-Tertiary Boundary in the Gulf of Mexico: A Restudy of DSDP Leg 77 Sites 536 and 540," *Geology* 20, no. 8 (August 1992): 697–700, https://doi.org/10.1130/0091-7613(1992)020%3C0697:PIDATC%3E2.3.CO;2.

7. Timothy J. Bralower, Charles K. Paull, and R. Mark Leckie, "The Cretaceous-Tertiary Boundary Cocktail: Chicxulub Impact Triggers Margin Collapse and Extensive Sediment Gravity Flows," *Geology* 26, no. 4 (1998): 331–334, https://doi.org/10.1130/0091-7613(1998)026%3C0331:TCTBCC%3E2.3.CO;2.

8. Lei Zhou, Frank T. Kyte, and Bruce F. Bohor, "Cretaceous/Tertiary Boundary of DSDP Site 596, South Pacific," *Geology* 19, no. 7 (July 1991): 694–697, https://doi.org/10.1130/0091-7613(1991)019<0694:CTBODS>2.3.CO;2.

9. J. Smit, "The Global Stratigraphy of the Cretaceous-Tertiary Boundary Impact Ejecta," *Annual Review of Earth and Planetary Sciences* 27 (May 1999): 75–113, https://doi.org/10.1146/annurev.earth.27.1.75.

10. Peter Schulte et al., "The Chicxulub Asteroid Impact and Mass Extinction at the Cretaceous-Paleogene Boundary," *Science* 327, no. 5970 (March 5, 2010): 1214–1218, https://doi.org/10.1126/science.1177265.

11. Christopher M. Lowery et al., "Ocean Drilling Perspectives on Meteorite Impacts," *Oceanography* 32, no. 1 (2019): 120–134, https://doi.org/10.5670/oceanog.2019.133.

12. Paul R. Renne et al., "Time Scales of Critical Events around the Cretaceous-Paleogene Boundary," *Science* 339, no. 6120 (2013): 684–687.

13. Robert A. DePalma et al., "A Seismically Induced Onshore Surge Deposit at the KPg Boundary, North Dakota," *Proceedings of the National Academy of Sciences* 116, no. 17 (April 1, 2019), https://doi.org/10.1073/pnas.1817407116.

14. William J. Broad and Kenneth Chang, "Fossil Site Reveals Day That Meteor Hit Earth and, Maybe, Wiped Out Dinosaurs," *New*

York Times, March 29, 2019, https://www.nytimes.com/2019/03/29/science/dinosaurs-extinction-asteroid.html.

15. Sean P. S. Gulick et al., "The First Day of the Cenozoic," *Proceedings of the National Academy of Sciences* 116, no. 39 (September 24, 2019): 19342–19351, https://doi.org/10.1073/pnas.1909479116.

16. Nicholas St. Fleur, "Drilling into the Chicxulub Crater, Ground Zero of the Dinosaur Extinction," *New York Times*, November 17, 2016, https://www.nytimes.com/2016/11/18/science/chicxulub-crater-dinosaur-extinction.html.

17. Thomas Sumner, "How a Ring of Mountains Forms inside a Crater," *Science News*, November 17, 2016, https://www.sciencenews.org/article/how-ring-mountains-forms-inside-crater.

18. G. S. Collins et al., "A Steeply-Inclined Trajectory for the Chicxulub Impact," *Nature Communications* 11 (2020), https://doi.org/10.1038/s41467-020-15269-x.

19. Gulick et al., "The First Day of the Cenozoic"; Ian Tiseo, "Global Historical CO_2 Emissions from Fossil Fuels and Industry 1750–2021," Statista, June 2, 2023, https://www.statista.com/statistics/264699/worldwide-co2-emissions/.

20. Owen B. Toon et al., "Environmental Perturbations Caused by the Impacts of Asteroids and Comets," *Reviews of Geophysics* 35, no. 1 (February 1997): 41–78, https://doi.org/10.1029/96RG03038.

21. Lucas Joel, "The Dinosaur-Killing Asteroid Acidified the Ocean in a Flash," *New York Times*, October 21, 2019, https://www.nytimes.com/2019/10/21/science/chicxulub-asteroid-ocean-acid.html.

22. Michael J. Henehan et al., "Rapid Ocean Acidification and Protracted Earth System Recovery Followed the End-Cretaceous Chicxulub Impact," *Proceedings of the National Academy of Sciences* 116, no. 45 (2019): 22500–22504, https://doi.org/10.1073/pnas.1905989116.

23. David A. Kring et al., "Probing the Hydrothermal System of the Chicxulub Impact Crater," *Science Advances* 6, no. 22 (2020), https://doi.org/10.1126/sciadv.aaz3053.

Chapter 14

1. Claude E. ZoBell and D. Quentin Anderson, "Vertical Distribution of Bacteria in Marine Sediments," *AAPG Bulletin* 20, no. 3 (March 1936): 258–269, https://doi.org/10.1306/3D932DB2-16B1-11D7-8645000102 C1865D; FengPing Wang et al., "Discovering the Roles of Subsurface Microorganisms: Progress and Future of Deep Biosphere Investigation," *Chinese Science Bulletin* 58 (2013): 456–467.

2. Beebe's amazing life is described in this fine biography: Carol Grant Gould, *The Remarkable Life of William Beebe: Explorer and Naturalist* (Washington, DC: Island Press, 2004).

3. Amanda Leigh Mascarelli, "Geomicrobiology: Low Life," *Nature* 459 (2009): 700–773, https://doi.org/10.1038/459770a.

4. John B. Corliss et al., "Submarine Thermal Springs on the Galápagos Rift," *Science* 203, no. 4385 (1979): 1073–1083, https://doi.org/10.1126/science.203.4385.1073.

5. Rachel Carson, *The Sea Around Us* (New York: Oxford University Press, 1951; New York: Open Road Media, 2011), 63. Citations refer to the Open Road Media edition.

6. Thomas Gold, "The Deep, Hot Biosphere," *Proceedings of the National Academy of Sciences* 89, no. 13 (1992): 6045–6049, https://doi.org/10.1073/pnas.89.13.6045.

7. Steven D'Hondt et al., "IODP Advances in the Understanding of Subseafloor Life," in "Scientific Ocean Drilling: Looking to the Future," special issue, *Oceanography* 32, no. 1 (2019): 198–207.

8. Ronald John Parkes et al., "Deep Bacterial Biosphere in Pacific Ocean Sediments," *Nature* 371, no. 6496 (1994): 413.

9. Steven D'Hondt et al., "Distributions of Microbial Activities in Deep Subseafloor Sediments," *Science* 306, no. 5705 (2004): 2216–2221, https://doi.org/10.1126/science.1101155.

10. Erwan G. Roussel et al., "Extending the Sub-Sea-Floor Biosphere," *Science* 320, no. 5879 (2008): 1046, https://doi.org/10.1126/science.1154545.

11. "Deep Sea Microbes Dormant for 100 Million Years Are Hungry and Ready to Multiply," Science Daily, July 28, 2020, https://www.sciencedaily.com/releases/2020/07/200728113533.htm.

12. Elizabeth Pennisi, "Scientists Pull Living Microbes, Possibly 100 Million Years Old, from Beneath the Sea," *Science | AAAS*, July 28, 2020, https://www.sciencemag.org/news/2020/07/scientists-pull-living-microbes-100-million-years-beneath-sea.

13. Felix Beulig et al., "Rapid Metabolism Fosters Microbial Survival in the Deep, Hot Subseafloor Biosphere," *Nature Communications* 13, no. 1 (2022): 1–9, https://doi.org/10.1038/s41467-021-27802-7.

14. William H. Horne et al., "Effects of Desiccation and Freezing on Microbial Ionizing Radiation Survivability: Considerations for Mars Sample Return," *Astrobiology* 22, no. 11 (November 1, 2022): 1337–1350, https://doi.org/10.1089/ast.2022.0065.

Chapter 15

1. Dennis E. Hayes et al., "Introduction," in *Initial Reports of the Deep Sea Drilling Project*, vol. 28 (Washington, DC: US Government Printing Office, 1975), https://doi.org/10.2973/dsdp.proc.28.1975.

2. Hayes et al.

3. James P. Kennett, "Cenozoic Evolution of Antarctic Glaciation, the Circum-Antarctic Ocean, and Their Impact on Global Paleoceanography," *Journal of Geophysical Research* 82, no. 27 (1977): 3843–3860, https://doi.org/10.1029/JC082i027p03843.

4. Carlota Escutia et al., "Keeping an Eye on Antarctic Ice Sheet Stability," *Oceanography* 32, no. 1 (2019): 32–46, https://doi.org/10.5670/oceanog.2019.117.

5. Escutia et al.

6. R. M. McKay et al., "Antarctic Cenozoic Climate History from Sedimentary Records: ANDRILL and Beyond," *Philosophical Transactions of the Royal Society A: Mathematical, Physical and Engineering Sciences* 374, no. 2059 (January 28, 2016): 20140301, https://doi.org/10.1098/rsta.2014.0301.

7. Escutia et al., "Keeping an Eye on Antarctic Ice Sheet Stability."

8. IPCC, "Summary for Policymakers," in *Climate Change 2021: The Physical Science Basis. Contribution of Working Group I to the Sixth Assessment Report of the Intergovernmental Panel on Climate Change* , ed. V. Masson-Delmotte et al. (Cambridge: Cambridge University Press, 2021), 3–32, doi:10.1017/9781009157896.001.

Chapter 16

1. J. P. Kennett and L. D. Stott, "Abrupt Deep-Sea Warming, Palaeoceanographic Changes and Benthic Extinctions at the End of the Palaeocene," *Nature* 353, no. 6341 (September 1991): 225–229, https://doi.org/10.1038/353225a0.

2. Francesca A. McInerney and Scott L. Wing, "The Paleocene-Eocene Thermal Maximum: A Perturbation of Carbon Cycle, Climate, and Biosphere with Implications for the Future," *Annual Review of Earth and Planetary Sciences* 39, no. 1 (April 2011): 489–516, https://10.1146/annurev-earth-040610-133431.

3. Thomas Westerhold et al., "Astronomical Calibration of the Paleocene Time," *Palaeogeography, Palaeoclimatology, Palaeoecology* 257, no. 4 (February 2008): 377–403, https://doi.org/10.1016/j.palaeo.2007.09.016.

4. Richard E. Zeebe and Lucas J. Lourens, "Solar System Chaos and the Paleocene–Eocene Boundary Age Constrained by Geology and Astronomy," *Science* 365, no. 6456 (August 30, 2019): 926–929, https://doi.org/10.1126/science.aax0612.

5. E. Anagnostou et al., "Proxy Evidence for State-Dependence of Climate Sensitivity in the Eocene Greenhouse," *Nature Communications* 11, no. 1 (September 7, 2020): 4436, https://doi.org/10.1038/s41467-020-17887-x.

6. McInerney and Wing, "The Paleocene-Eocene Thermal Maximum."

7. McInerney and Wing.

8. James C. Zachos et al., "Tempo and Scale of Late Paleocene and Early Eocene Carbon Isotope Cycles: Implications for the Origin of

Hyperthermals," *Earth and Planetary Science Letters* 299, no. 1–2 (October 15, 2010): 242–249, https://doi.org/10.1016/j.epsl.2010.09.004.

9. Victor A. Piedrahita et al., "Orbital Phasing of the Paleocene-Eocene Thermal Maximum," *Earth and Planetary Science Letters* 598 (November 15, 2022): 117839, https://doi.org/10.1016/j.epsl.2022.117839.

10. Philip D. Gingerich, "Temporal Scaling of Carbon Emission and Accumulation Rates: Modern Anthropogenic Emissions Compared to Estimates of PETM Onset Accumulation," *Paleoceanography and Paleoclimatology* 34, no. 3 (March 2019): 329–335, https://doi.org/10.1029/2018PA003379.

Chapter 17

1. Charles Darwin, *On the Origin of Species* (London: John Murray, 1859), 42.

2. Charles Darwin, *The Descent of Man, and Selection in Relation to Sex* (London: John Murray, 1871), 3.

3. Peter B. deMenocal, "Climate Shocks," *Scientific American* 311, no. 3 (2014): 48–53, https://doi.org/10.1038/scientificamerican0914-48.

4. Elisabeth S. Vrba et al., *Paleoclimate and Evolution, with Emphasis on Human Origins* (New Haven, CT: Yale University Press, 1995).

5. Raymond A. Dart, "*Australopithecus africanus*: The Man-Ape of South Africa," *Nature* 115, no. 2884 (February 1925): 195–199, https://doi.org/10.1038/115195a0.

6. Dart, 199.

7. Robert Ardrey, *The Territorial Imperative: A Personal Inquiry into the Animal Origins of Property and Nations*, A Delta Book (New York: Atheneum, 1966), https://books.google.com/books?id=xJhqAAAAMAAJ.

8. DeMenocal, "Climate Shocks," 48–53.

9. DeMenocal.

10. Sarah J. Feakins et al., "Northeast African Vegetation Change over 12 M.Y.," *Geology* 41, no. 3 (2013): 295–298, https://doi.org/10.1130/G33845.1.

11. Richard Potts and Christopher Sloan, "Survival of the Adaptable," in *What Does It Mean to Be Human?* (Washington, DC: National Geographic, 2010), 44–53.

12. Richard Potts and J. Tyler Faith, "Alternating High and Low Climate Variability: The Context of Natural Selection and Speciation in Plio-Pleistocene Hominin Evolution," *Journal of Human Evolution* 87 (2015): 5–20, https://doi.org/10.1016/j.jhevol.2015.06.014.

13. Potts and Sloan, "Survival of the Adaptable," 49.

Chapter 18

1. Marcia K. McNutt, "Achievements in Marine Geology and Geophysics," in *50 Years of Ocean Discovery: National Science Foundation 1950–2000* (Washington, DC: National Academies Press, 2000), 63, https://nap.nationalacademies.org/read/9702/chapter/6.

2. National Research Council, *Scientific Ocean Drilling: Accomplishments and Challenges* (Washington, DC: National Academies Press, 2011), https://doi.org/10.17226/13232.

3. IODP, "2050 Science Framework," International Ocean Discovery Program, accessed June 13, 2023, https://www.iodp.org/2050-science-framework.

4. Vannevar Bush, *Science, the Endless Frontier; a Report to the President on a Program for Postwar Scientific Research* (Washington, DC: US Government Printing Office, 1945), 11.

Index

Note: Page numbers in italics refer to figures.